Design of DNA, Genetic Codes, and Life Function

Charles D. Schaper, Ph.D.

Molecular Sciences Publishing House

=======

Version 1.0: July 4, 2020

Schaper, Charles
Design of DNA, Genetic Codes, and Life Function
Total: 145 pages, 75 figures and tables
Includes encoding, coding, decoding, and translation tables
ISBN 978-1-7353721-3-6
I. Book design II. Title and Contents

Molecular Sciences Publishing House
a division of Technology Bin LLC
Union City, CA, 94587
United States of America
cschaper@transferdevices.com

Contents

Biographical Sketch

Relevant to this subject matter, Charles Schaper's background includes training in engineering, chemistry, information theory, and nanotechnology, as follows:

- B.S. and M.S degrees in Chemical Engineering from the University of Connecticut, including a master's thesis with Professor Cooper on state-space design of control systems.
- Ph.D. in Chemical Engineering from the University of California, Santa Barbara with co-advisors Professors Seborg and Mellichamp, including a dissertation on stochastic robust control theory.
- Post-Doctorate research at the Information Systems Laboratory, Electrical Engineering Department, Stanford University working with Professors Kailath, Boyd, Franklin and Saraswat on applications of communications, computation, control and signal processing to silicon semiconductor manufacturing.
- At Stanford University spanning fifteen years as academic research staff, electrical engineering, ultimately as senior research scientist working with Professors Kailath and Pease, which included a patented invention for the replication and transfer of microstructures and nanostructures.
- Founder and president of a nanotechnology start-up company on molecular transfer lithography, which included working as Principal Investigator on National Science Foundation grants.
- Admitted to medical school, M.D. program, for the 2019/2020 academic class at St. James Medical School, although have not started coursework yet, just external preliminary classes, with plans to defer the official start for another year.

Preprints

The path for the results described herein is documented by the following set of preprints, which is listed in reverse chronological order. It began from a mathematical investigation of a fever in response to exercise after vaccination, which was conducted as a personal curiosity, as are all the subsequent results. After about a six month break from the first two preprints, I investigated the underlying chemistry of fever first at the systems level, then the cellular level, and finally the nuclear level in continuing to search for the fundamental reasons, that led to the results described herein on DNA structure and function.

1. Schaper, Charles (2020): Intermolecular Enzymatic Encoding of Nucleic Acid, Steroid Complexes: A New Theory on the Chemical Origin of Life Based on Evidence of Structural Symmetry. https://arxiv.org/abs/2006.06429.

2. Schaper, Charles (2020): Intermolecular Binding of Functional Cyclic Compounds to RNA Nucleotides by an Ionic Process. Preprints.org. https://doi.org/10.20944/preprints202004.0522.v1

3. Schaper, Charles (2020): A Mechanism of Action for Hydroxychloroquine and Azithromycin to Inhibit Coronavirus Disease COVID-19. ChemRxix. https://doi.org/10.26434/chemrxiv.12148914.v1

4. Schaper, Charles (2020): New Code for DNA Nucleotide Sequences. ChemRxiv. https://doi.org/10.26434/chemrxiv.12003786.v1

5. Schaper, Charles (2020): Structural Symmetry of DNA Nucleotides and Steroid Hormones. ChemRxiv. https://doi.org/10.26434/chemrxiv.11991567.v1

6. Schaper, Charles (2020): Endogenous Binding of Steroid Molecules to DNA Nucleotides by a Ca2+/PO4- Process to Enable Gene Transcription. ChemRxiv. https://doi.org/10.26434/chemrxiv.11868261.v2

7. Schaper, Charles (2020): Molecular Pathways Resultant in Fever due to Inhibitory Effects of Binding PGE2 to Glucocorticoid Receptors. BioRxiv. https://www.biorxiv.org/content/10.1101/2020.01.14.907105v1

8. Schaper, Charles (2019): Competitive Inhibition of Cortisol by Prostaglandins at the Ligand Binding Domain of Glucocorticoid Receptors. BioRxiv. https://www.biorxiv.org/content/10.1101/851501v7

9. Schaper, Charles (2019): Identifying Integrative Molecular Pathways for Predictive Modeling of Infectious Disease. BioRxiv. https://www.biorxiv.org/content/10.1101/835546v1

10. Schaper, Charles (2019): Analytic Model of fMRI BOLD Signals for Separable Metrics of Neural and Metabolic Activity. BioRxiv. https://www.biorxiv.org/content/10.1101/573006v1

11. Schaper, Charles (2019): Dynamics and Internal Control of Body Temperature in Response to Infectious Agents and other Causal Factors. BioRxiv. https://www.biorxiv.org/content/10.1101/566679v1

Provisional Patents

Charles Schaper has filed several personal provisional patents associated with this work, of which the application in approximate form was subsequently published on preprint servers. These provisional patents comprise the following set including one that covers the material described herein, submitted for review under confidential disclosure.

1. Charles Schaper, "Binding Steroid Molecules to DNA", Provisional Patent: US 62/977,216. Filed 3/9/2020.
2. Charles Schaper, "Binding Nucleic Acids", Provisional Patent: US 63/016,446. Filed 5/5/2020.
3. Charles Schaper, "Synthesis of Genetic Structures", Provisional Patent: US 63/027,320. Filed 5/19/2020.
4. Charles Schaper, "Design of DNA, Genetic Codes, and Life Function", Provisional Patent. Ref: E20206R233375951. Filed 6/28/2020.

Preface

The origin of life is a great unsolved mystery, and in this manuscript, a comprehensive solution is presented for the first time. Focusing on the information content and its access, this manuscript covers encoding, transmission, decoding and translation of DNA, including the identification of the origin of DNA and its mechanism of action to produce proteins from amino acids immediately after its synthesis. Coding and decoding tables are provided, as is the molecular basis for the structure and function of DNA in the form of a primer molecule. Further replication, error correction, and genetic rearrangement processes are presented. As described in the manuscript, DNA is co-synthesized with the steroid molecule in a co-enzymatic complex, and thus the work is also fundamental in describing the structure and function of steroid molecules relative to DNA molecules, as both are required to induce life function. The results are enabled from multiple discoveries that I have made and discuss herein. Furthermore, the construct is described of a digital framework that results from molecular structures and intermolecular coupling arrangements.

I trust this solution to be accurate and correct as it is consistent with observable metrics, answering many detailed questions on DNA and its function that prior to this work were without an answer. The results are described in the format of a first draft, and thus another edit is underway, but all of the main results are presented in cohesive format permitting a new understanding of DNA, as well as its interaction with steroid molecules. Together, instantaneous life function is the result at the completion of the synthesis comprising the manufacture of a molecular structured communication path, enabling simultaneous structure and genetic function. The results are powerful, illuminating, and should impact along multiple levels of science, medicine and philosophy.

Chapter 1

Introduction

The origin of life is inherent to central tenets of science and philosophy; thus, as it remains one of the greatest unsolved mysteries, many prospective theories have been developed and studied at great length [1, 2, 3]. To focus this broad topic, as all cellular organisms of life contain DNA, the origin of life is synonymous with revealing the mechanisms responsible for synthesizing DNA starting from basic materials and processes that would have come into existence about the time Earth was formed, over four billion years ago [4, 5]. In terms of the creation of the first DNA molecule, there are excellent articles written on the range of ideas, such as those that revolve around starting with RNA molecules and others that are based upon geological events [6, 7, 8, 9]. Thus, in putting forth the solution for the origins of both the structure and function of DNA, particularly as it relates to its information content and distribution, as this exposition develops, it would be distractive to compare the new approach directly with all of the alternative developments, because of the wide range, and varying levels of evidentiary basis [10, 11]. Therefore, in this introductory section, the objective is to characterize the constituents of a valid approach that reveals the synthesis of life function. For each requirement, the results of my analysis are highlighted that characterize new developments in substantiating a solution to the design of DNA and its genetic code.

In developing a quantitative approach on the origin of life, a difficult problem to address immediately is the so-called "chicken and egg dilemma", as to which one came first [12]. For example, in the case of the transcription process under the current approaches, DNA is required to form proteins, but proteins are needed by DNA to initiate transcription. Thus, there has been the speculation that DNA came about from an "RNA world", and this theory has taken the preeminent position in explaining the origin of life [13]. Thus, this approach would generate DNA from building blocks, but it would be unclear as to the driving force behind this self-assemble process of significant complexity. In the developments presented in this research, I show that the answer to the question of which one came first, the chicken or the egg, the answer is that both were synthesized together within a common enzymatic framework, developed in unison: that is, the DNA molecule, as well as steroid molecules that are needed for DNA to function, are developed in an intermolecular linked manner at the same time in tandem with the steroid molecule affecting the formation of the DNA molecule and vice versa, both starting from the same molecule type, paired together and co-synthesized in a single unified complex.

A valid solution to the origin of DNA should also provide answers to detailed questions on the molecular structure of DNA. For example, detailed structural questions have to be answered: why are purine and pyrimidine the bases that comprise DNA? How are the dimensions established between pairs of nucleotides? Why does adenine-thymine only have two internal hydrogen bonds, whereas cytosine-guanine have three internal hydrogen bonds? How is the synthesis of DNA related to the formation of the double helix? What is the structural basis for DNA nucleotides? Many theories on the origin of life never even get to this level of detail to provide a reasonable response to these basic questions, and most suggest that it is this way, because it has to be this way. In these developments, however, a process sequence is developed that within its first few steps provides quantitative answers to such structural questions on DNA.

Another important requirement in the determining the origin of DNA is that the materials and processes utilized in forming DNA should be feasible during the time when Earth was formed. Some have considered DNA to have been possibly constructed up to 4 billion years ago, which is close to the estimates on the time at which Earth was formed [14]. To accommodate for further pushbacks, theories should even consider DNA to be generated essentially at the same time as the formation of the earth. Thus, lengthy evolutionary processes, or theories based upon chance of multiple sequential events that have low yields, would be hard to satisfy this category. The materials and processes should be developed that are consistent with these conditions, and the processes should be relatively rapid, and preferably simple. Most theories on the origin of DNA are valid in applying chemicals and processes that would have been present at the time when Earth was formed, but have challenges in terms of its simplicity. However, as shown here, the materials and process steps are consistent with these conditions, and the processes are of an elementary nature defined at the onset of the synthesis.

Ideally, the answer to the origin of DNA should not only explain the past, but offer a path going forward. Approaches which utilize actions by chance or require lengthy development periods are less interesting than those that can be used in synthesis applications or to discover new medical knowledge. DNA is central to the operation, and a better understanding of it should result in new medicines, new therapies, and new approaches. Random events are thus not satisfying in the development of pursuits for the creation of life function through DNA, and are not helpful in addressing basic human inquiries on the origins of life. Here, I show that DNA is the result from the synthesis of a logical set of steps, which can lead not only to a better understanding of DNA, but also to produce new ideas such as in the transcription process itself.

One great challenge in the development of a solution on the origin of DNA is in testing the approach, which may imply the capability to recreate it. That is a daunting challenge as for some approaches as it may not be possible to develop a protocol to test the hypothesis, especially those that rely on slow periods of evolution, random events, or exogenous processes. However, in this work, there is a chemical structure embedded within DNA, which has underlying characteristic of the structure of DNA, that makes it possible to have a first line of comparative analysis. In this book, the identification of a discovery of the structural basis of DNA is identified for the first time, in which it is indicated that the molecular structure of the steroid molecule is embedded into each and every DNA nucleotide to the present day. Thus, the solution developed here is matched to this discovery, thereby providing a verifiable metric of its validity, which the alternative approaches do not address. Moreover, it is further shown through analysis of measurable efficacy of certain pharmaceuticals, that the results are consistent with the structural arrangements involved in the molecules comprising the strategy of synthesizing DNA and steroid molecules by a basic set of processes, and hence provide a biomarker as evidence of the developed solution.

Most importantly and generally the largest stumbling block is the resolution of the "code of life", the communication channels to generate information as encoding, transmission, and decoding. Literature acknowledges that this is the biggest problem for the current theories [15, 16, 17]. But here I show a clear answer and comprehensive solution in explaining the code of life as AGTC as well as its relationship with amino acids and translation to proteins. Thus, while the other approaches, such as the RNA world, describe how the molecules can be formed, they do not characterize the organizational strategy associated with the capability of DNA to store and retrieve information. In the second step of the process flow, it is shown how the code of life is generated and its functional nature established. Combined with the alignment of the results with experimental evidence including those of a structural and functional basis of DNA, these are significant reasons to support the validity of the developed solution as the preemptive description on the origin of DNA.

Finally, this new solution on the origin of DNA introduces another new concept that has never been shown before: the concept of synthesizing the mechanisms of the structure and function of DNA simultaneously in a unified complex, in which structure and function are effectively designed as one. Thus, when the co-formative processes are complete, DNA and its associated steroid molecules, along with the amino acid mapping to base pairs, are put into practice immediately to establish in an instant a central life function of replication and protein translation through a transcription procedure. Therefore, the equivalent of a big bang moment for life function is enabled, of which there is paired synthesis of DNA and steroid molecules, developed in tandem as a co-synthesized intermolecular unified complex, dependent upon each other in one sequence of steps, co-optimized in structure and function, released simultaneously.

Furthermore, since the purpose of DNA is to translate to proteins, a valid solution on its origin needs to describe the mechanism by which the complementary base pairs map to amino acids. Again, almost all alternative approaches do not address this issue as they do not even have the capability to determine how the base pair code is defined in the first place, and thus leave this important component unattended. However, here I describe how this mapping of DNA base pair to amino acid is achieved indicating that it is a natural outcome of the procedure of associating the integrated, paired synthesis also permits encoding of an intermediary to achieve the amino acid to base mapping. The information channel is established of an encoder, a transmission mechanism, and a decoder, as well as translation.

Although there are challenges in developing a solution to provide insight on the creation of DNA, it is possible to develop a checklist for the requirements associated with an explanation on the origin of life. In so doing, exciting explanations of the past are possible, unraveling many mysteries, and a path going forward is developed for exciting new discoveries. Therefore, as this solution that will be described herein satisfies this checklist of basic requirements, it should enable this approach as a leading candidate among all theories describing the synthesis of the structure and function of DNA, its molecular triggers, that is specialized molecules of a steroid construct, for protein production, and hence enable a chemical origin of life.

1.1 Historical context

With the discovery of a new solution described herein on the design of DNA and genetic codes, it is interesting to consider the results in context with the historical discoveries of DNA to determine at which point the new results fit. The molecules of DNA were formulated as far back as the 1800's of which DNA was discoveredby

Friedrich Miescher, who isolated the molecule, later known as DNA, from a cell nucleus. Shortly afterwards, Albrect Kossel identified the nucleic acid as the isolated molecule, and isolate the nitrogen bases of Adenine, Cytosine, Guanine, Thymine, in DNA and Uracil in RNA. Schrodinger's aperiodic crystal [18] indicated a structure associated with the origin of life, followed by Watson and Crick [19] who discovered the structure of DNA as the double helix in 1953. However, subsequent to that event, there was speculation on the function and how it was generated. The original molecular biologists favored DNA as the origin of life [20] but had no clear answer on how that came to be, nor how the genetic code was derived. Thus, numerous theories were developed to try to explain the origins of DNA, but known successful to include the genetic codes, and some could be classified as ad hoc hypotheses.

Therefore, as this book describes new results by taking a different route of anlysis of DNA, starts from the moment in 1953 when the double helix was discovered with the base pairings of A-T and C-G. The basic difference in my approach begins from a structural analysis of the findings of Watson and Crick. What I show in Chapter sec:discovery is that the DNA nucleotides comprising the double helix have an embedded molecular structure that acts as a primer and clue as to the origins of DNA structure as well as its function. What the original discoveries of DNA failed to consider was that DNA did not derive in and of itself, but as I show DNA was actually part of a larger complex that contained not only DNA but also the capacity to make DNA function. Thus, herein, I aim to prove that the original DNA molecule was but half of a co-enzymatic, co-synthesized, co-encoded "master" complex, in which the other half was comprised of ultimately its triggering elements, that is steroid molecules, which enabled life function pursuant to DNA when both were simultaneously released in a decomposition of the master complex. The details will follow on how this fits together, and is pursuant to Occam's Razor, in that as soon as the master complex is synthesized, and separated, life function is created, and the genetic processes can begin, that is, the originating work is done with sufficient concentration of the released molecules to permit further development immediately.

1.2 Advancements and consistencies

Many discoveries are made in taking a different route in the analysis of the DNA structure, beginning with the identification of a structural basis of the DNA molecule first described sixty-seven years ago. It is thus a fundamental finding, and apparently difficult to uncover as it has not been described over such a long period, until now, as first described in patent applications and in the preprints [21, 22, 23]. This work leads to the first description of a table of translation mapping one set of molecular structures of DNA nucleotides and its associated steroid molecules, to a second set of structures, the set of amino acids comprising proteins as presently known, described in Chapter 6. Moreover, a molecular structure and process is indicated which establishes a digital communication framework from chemistry. This may be one of the first descriptions of such a digitalframework created through molecular physical chemistry.

Basic discoveries enabling the work are: (1) the discovery of a primer in the basic structure of the DNA nucleotide pairs as the trace of a steroid molecule. It was very surprising to me that even though the structure of DNA was discovered more than 50 years ago that no one has reported on this embedded structure. I described the findings first in patent applications and in the preprint. Thus it indicated that to me that steroid molecules have a major influence over DNA, and indeed they do; (2) the discovery of the missing hydrogen bond of the A-T pair relative to C-G. When paired with corticosteroids, A-T then has three hydrogen bonds to match C-G. This requirement enabled the unveiling of the code of DNA. Thus, this work also defines two classes of steroids. Current work will map this result to the two types of responses seen in human physiology. (3) Besides the structure of the base 4 DNA code, the third discovery is an interaction vessel that links DNA nucleotides, steroid molecules, and amino acids, that thereby defines the nucleotide sequence as a triplet of DNA nucleotides to couple directly with amino acids. Many other discoveries are made along the way, which include the unified complex defining the origin of DNA; its simultaneous release to separate the DNA double helix from its functional agents, the steroid molecules; the genetic rearrangement strategies to produce variety of protein structure and DNA sequences; the replication and error correction procedures. Keep in mind that these developments are consistent with the experimental evidence of the steroid molecular structure embedded into each and every DNA nucleotide pair.

Chapter 2

Discovering a Molecular Primer of DNA

The encoding, transmission, decoding, and translation mechanisms for DNA were enabled by a discovery that I made at the initiation of my investigatory path associated with DNA research. The fundamental result was initially described in my preprints, [21, 22, 23], after filing for provisional patents. These developments are critical to understand as it provides an experimental basis at the onset of the investigation, and also provides the basic concept that enables the design of an enzymatic complex that encodes molecular information systems.

To develop the new solution on the origin of DNA structure and function, the experimental evidence consistent with the solution is presented along three lines, two of which constitute discoveries and the third is confirming evidence. The first comprises a structural analysis of DNA to illustrate that there is a common shape embedded within each of the four DNA nucleotides, which will serve as a beacon to develop the new solution. It is in the shape of a steroid molecule and the match is perfect along seventeen elements. Next, a functional analysis of DNA is performed, which was a discovery that a perfect intermolecular link to corticosteroids provide the missing third hydrogen bond of the A-T base pairing relative to the G-C base pairing. Thus, the interaction with steroids relates to both structure and function. This is further confirmed by analyzing pharmacological data associated with the corticosteroids to indicate that efficacy is matched to the capability to form an intermolecular hydrogen bond at a precise position with thymine.

2.1 Structural symmetry of steroid molecules and DNA nucleotides

To develop an experimental basis to initiate the approach, and thus providing a source of confirmation at the onset of the developments, a structural analysis of DNA nucleotide base pairings is analyzed. In Figure 2.1(a), the front of the DNA nucleotides is indicated for each of the four pairings as listed from top to bottom in the figure as Adenine-Thymine (A-T), Cytosine-Guanine (C-G), Thymine-Adenine (T-A), and Guanine-Cytosine (G-C). In Figure 2.1(b), the backside of each of the four DNA nucleotide pairings is illustrated. This is obtained by flipping the front orientation in a horizontal direction, and both sides are important to the analysis. Also included in the illustrations are the phosphodiester backbone and the sugar molecules. The arrangements are envisioned as the DNA nucleotide structure printed on paper, and viewing it from the front and the back side. Also included in the illustration is the hydrogen bonding of the base pairings, in which it is shown that there are three hydrogen bonds for the C-G and G-C pairs, and two internal hydrogen bonds for the A-T and T-A pairs. It is of importance that at the carbon two position of thymine, there exists a functional oxygen element, but there is no corresponding element on adenine that could provide a hydrogen bond as a link.

To indicate that the four fused ring structure of a steroid is a basis for DNA, the extraordinary result is presented in the four red overlays each of Figure 2.2(a) and (b). This overlay is obtained by connecting precisely a total of seventeen oxygen, nitrogen, and carbon elements of the base pairs in a particular manner to produce a tetra-ring structure embedded within each of the four DNA nucleotide pairs. This can be achieved for each base pair, front and back, and the match is in perfect correspondence when comparing each of the matches. The four-ring structure is exactly in the shape of a steroid molecule, which was originally identified in the preprints [21, 22]. The orientation of the molecule is in perfect correspondence when analyzing the frontside for the A-T and G-C pairings, where a purine molecule starts the association on the 5' side, while on the backside the alignment is perfect for the T-A and C-G pairings when the pyrimidine molecule starts the 5' pair. There is a stretching along one of the rings where the purine and pyrimidine come into alignment through hydrogen bonding. Thus, this association is a basis for DNA of which approaches to explain the origin of DNA must identify. The concepts developed here begins from the steroid molecule, and thus is in perfect harmony with a structural and functional basis of DNA.

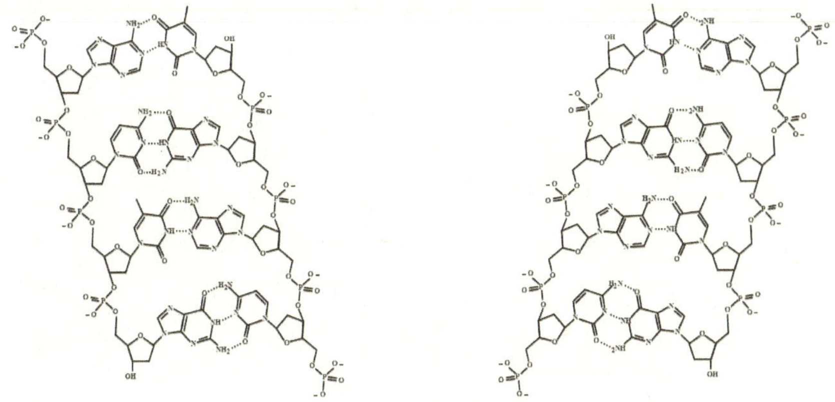

(a) Front (b) Back

Figure 2.1: (a) The front side of DNA with the base pairing starting from the top, A-T; C-G; T-A; G-C; (b) Examining the "back" side of the normal presentation, which is equivalent to that which would be seen on the back of a printed page with the molecule on the front, the common structure of the steroid molecule is obtained for each of the nucleotide pairs. The coupling of back side to front side is achieved by flipping the common structure, thus indicative that one common structure is appropriate for each of the pairs as the backside orientation becomes reoriented properly on the frontside, and vice versa.

2.2 Intermolecular binding of corticosteroids to thymine

The second line of support involves the symmetry of the steroid hormone with the DNA nucleotide pairs with respect to the intermolecular binding. Namely the corticosteroids, which have a mid-molecule group, should bind onto thymine. This result is consistent with the hydrogen bond that was presented very early into the development, essentially the definition of the DNA nucleotide code itself. It should be the case that the symmetry of the steroid molecule is capable of forming an intermolecular hydrogen bond with A-T and T-A, and not needed for C-G and G-C, the result is evaluated with the representation of DNA and the steroid hormones. As shown in Figures 2.3(a) and 2.3(b), this symmetry is indicated and shown. Of course, having developed the symmetry relations between steroid hormones and DNA nucleotides as described in the preprint [22], this mapping was an objective. Thus, it is apparent that this representation is still conserved to this date, which includes the relationship with steroid hormones, the orientation, the intermolecular bonds, the connection to the phosphodiester backbone, and even the configuration of the end groups. More-

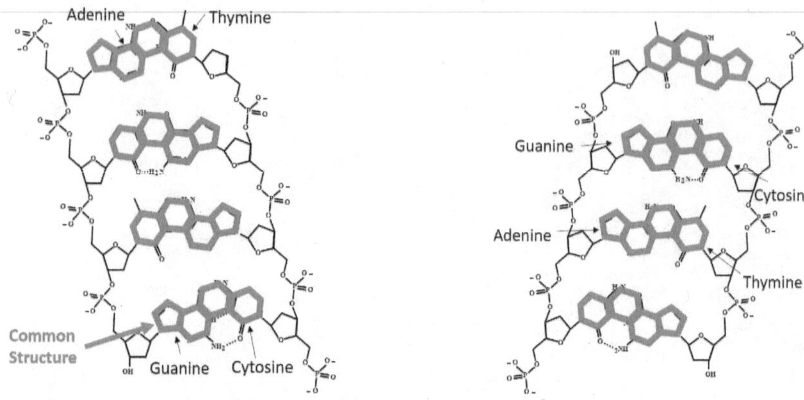

(a) Front (b) Back

Figure 2.2: (a) In a structural analysis of the four possible DNA nucleotide pairs presented from top to bottom as Adenine-Thymine (A-T), Cytosine-Guanine (C-G), Thymine-Adenine (T-A), and Guanine-Cytosine (G-C), there is a common structure in the shape of a four ring steroid molecule at each of the nucleotides.; (b) For the back side, it indicates that one common structure is appropriate for each of the pairs as the backside orientation becomes reoriented properly on the frontside, and vice versa.

over, together, the orientation and class of steroid hormone correspondence form a new code for DNA as shown in Figure 2.3(c), which is further described in the preprint [23]. Moreover, this is consistent with the indication in Section 3 which noted that the nucleotide pairs with pyrimidines along left (5') side had to be flipped in order to achieve their proper orientation consistent with the double helix. Hence it is logical that the steroid representation would involve a forward side and a reverse side of orientation. This configuration of forward and reverse orientation is thus apparently still conserved to this date in the DNA molecule. Thus, the second line of experimental evidence of a structural biology nature to identify a functional match to the steroid hormone, is in support of the new approach on the origin of life is successful of which steroid molecules and DNA nucleotides are synthesized together within the same unified complex.

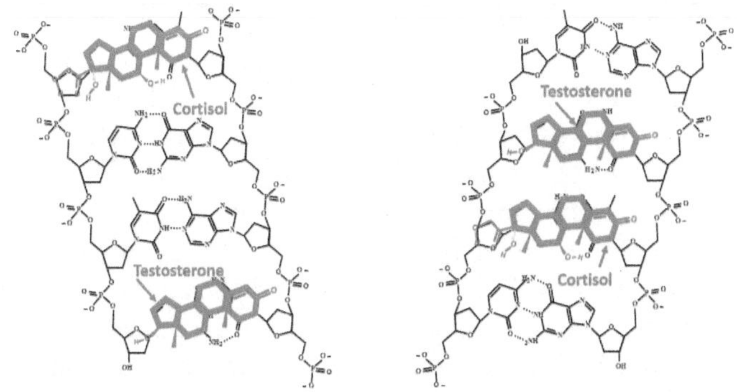

(a) Front with Steroid Hormone

(b) Back with Steroid Hormone

Figure 2.3: Consistent with the result of a common steroid molecule embedded in the DNA nucleotides, the intermolecular bonding capability is examined with respect to thymine. Classes of steroid molecules are defined with cortisol defining those molecules which have a mid-molecule functional group, and testosterone representing those that do not have a mid-molecule functional group. The objective is to overlay the steroid such that three hydrogen bonds are formed. on the common steroid structure for each of the four possible steroid structures, a class of steroid hormones are assigned with the objective of achieving a total of three hydrogen bonds when overlaid with the steroid hormone. In (a), the cortisol steroid hormone forms an intermolecular bond with the (A-T) pairing, and thus a total of three hydrogen bonds result. In the case for (G-C), there is already three internal hydrogen bonds, and thus testosterone is used as the overlay since it does not have a mid-molecule hydroxyl group. For cortisol and testosterone, the end groups are available for connection with the phosphodiester backbone, which was shown in the simulation. (b) Following the same protocol, cortisol forms an intermolecular hydrogen bond with the (T-A) pair, while testosterone couples with the (C-G) nucleotide pair, but on the back-side. This is consistent with the need to rotate the pyrimidine molecules at the 5' strand to achieve the double helix when ejecting the steroid triggering molecules.

2.3 Structural symmetry correlated with pharmacological efficacy

To further validate the method, experimental data is evaluated that can be used to prove the claim of this approach that the intermolecular bond which essentially established the DNA code is critical, and thus serves as a biomarker of DNA to guide in the development of an answer of its origin. In this experimental analysis, the intermolecular hydrogen bond is examined of coupling steroid hormones similar to cortisol to thymine for corticosteroid data. The handbooks indicate that the relative activity of the steroids dexamethasone:prednisolone:cortisol is approximately 25:4:1 and that prednisone, in its native form, is inactive as it must be converted into prednisolone in the liver before it can be used [24]. This relationship is consistent with the intermolecular bond to thymine by its steroid triggering molecule as presented in this study. In Figure 2.4(a), the relationship of prednisolone to thymine is indicated, in which it is shown that it is possible to form an intermolecular hydrogen bond with thymine. However, in Figure 2.4(b), the relationship of prednisone to thymine is indicated, in which it is not possible to form an intermolecular hydrogen bond. Note that the OH group is the only difference between prednisolone and prednisone. This precise result is consistent with the efficacy of prednisone, which experimentally shows no activity, and must be converted into prednisolone before becoming active.

Further experimental data is consistent with the approach described herein. In Figure 2.4(c), the relationship between the steroid hormone cortisol and thymine is indicated, which can form an intermolecular hydrogen bond. With respect to prednisolone, the only difference is an extra double bond in the ring of prednisolone within thymine. The additional double bond will aid in stabilizing the intermolecular hydrogen bond of thymine and prednisolone, and may also provide favorable structural arrangement in terms of the angle. Further, by examination of dexamethasone in Figure 2.4(d), the activity will be even further enhanced, due to the fluoride group stabilizing the intermolecular hydrogen bond even more than prednisolone, thereby increasing its relative activity. Thus, the third line of evidence based on experimental data is consistent with the paired development of the steroid molecule and DNA nucleotides.

2.4 The primer

My discovery of both the structural basis for DNA as the steroid molecule, and the location for the missing third hydrogen bond of adenine-thymine as found in the corticosteroids, enables a guide for further discovery of the origins of DNA and its life function. This synthesis strategy can then be used as a guide for the development of similar molecular systems. The primer used in the development of the synthesis strategy, and its accompanying encoding scheme are as follows: (1) the steroid molecule is the basis for the structure of DNA as purine and pyrimidine will form 3 of the four rings, and its integration through hydrogen bonding, will form the fourth ring; (2) the steroid molecule and the DNA molecule are coupled and capable of

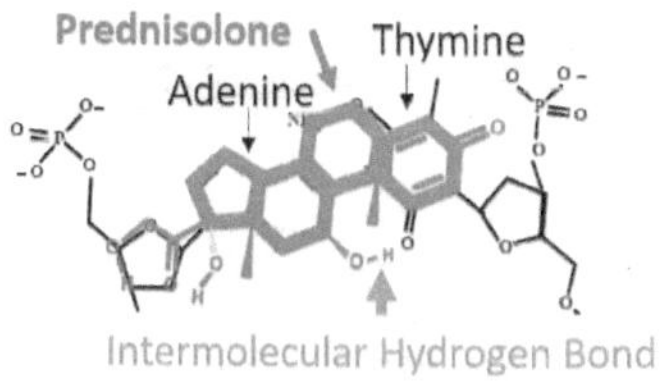

(a) Prednisolone

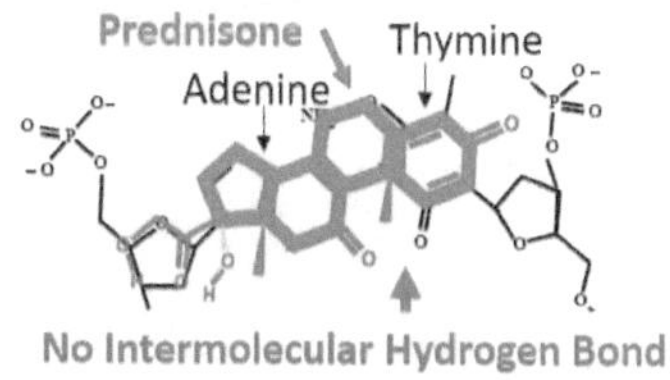

(b) Prednisone

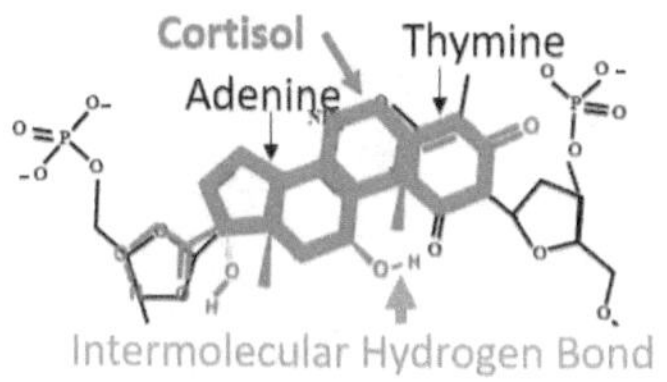

(c) Cortisol

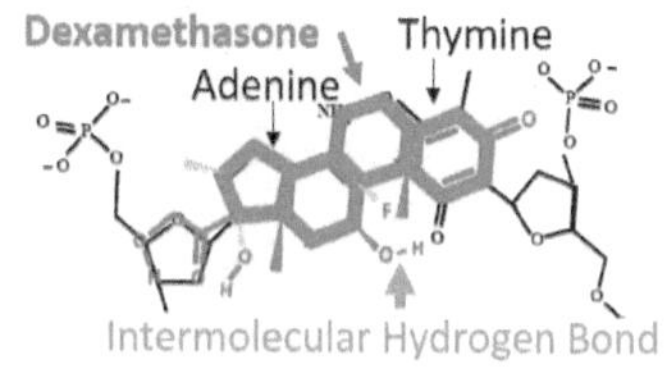

(d) Dexamethasone

Figure 2.4: For analysis of experimental results, the efficacy of steroid therapeutics is examined. (a) The capability to form an intermolecular hydrogen bond between prednisolone and thymine. (b) It is not possible to form an intermolecular hydrogen bond between prednisone and thymine. Indeed prednisone is inactive in practice, and must be converted to prednisolone in order for it to be potent. This result provides evidence in proving the claims of this manuscript. (c) The overlay of the nominal steroid hormone cortisol is shown capability to form an intermolecular hydrogen bond with thymine. With respect to prednisolone, the double bond in the ring adjacent to thymine aid in enhancing the intermolecular hydrogen bond strength, and thus increased potency is expected for prednisolone relative to cortisol, which is indeed the case in practice. (d) Dexamethasone also is capable of forming an intermolecular hydrogen bond with thymine. In addition, the fluoride group and double bond will aid in the potency, and thus it would be expected to have greater potency than prednisolone, and cortisol, which again aligns with that observed in practice

mirroring its each others actions by aligning the ring structures; (3) intermolecular hydrogen bonding plays an important role in establishing molecule to molecule interactions and functionality. Thus, communication between the steroid molecule and the DNA molecule, which have a common shape, is established through a structural organization, that is the four ring system, and a chemical organization, that is intermolecular hydrogen bonding. Understand these concepts very well because it will be applied many times in the following chapters.

Chapter 3

Molecular Complex for Encoding DNA

Representing information content as a molecular configuration requires a synthesis strategy that considers both the structure of the molecule as well as its functional characteristics. It is not merely enough to design the shape of the molecule, it is also required that the function, such as strand separation and translation, be taken into account during the design phase. If it is done correctly, then the entire procedure of both structure and function is set forth in motion, and thus will arrive automatically at the desired result of a self-communicating system capable of encoding, transmission, decoding, and translation into purpose. Thus, the encoding step as described in this section is critical. And it is also instructive as these initial coding steps and material selections are also consistent with the origin of DNA.

3.1 The primer molecule as the starting material

To select a starting material for the design of DNA, the primer molecule identified in the discovery described in Chapter 2, which is a steroid molecule comprised of four fused rings, with three fused rings of six-carbons, and one fused ring of five carbons. In Appendix A.1.5, the synthesis of this starting material using chrysene, which is comprised of four fused rings of benzene. In Figure 3.1(a), the basic starting material of chrysene is indicated, which can be used to generate the primer molecule of this study, indicated in Figure 3.1(b).

Of importance to these developments is the ring structure of the primer molecule and the relative arrangement of the ring structure when brought together. In the two formats indicated in Figure 3.2(a) and 3.2(b), the rings of the primer molecule of Figure 3.1(b) are solely indicated. From the five-carbon ring the orientation of the molecule can be determined. In Figure 3.2(a), the rings are color-coded. In Figure

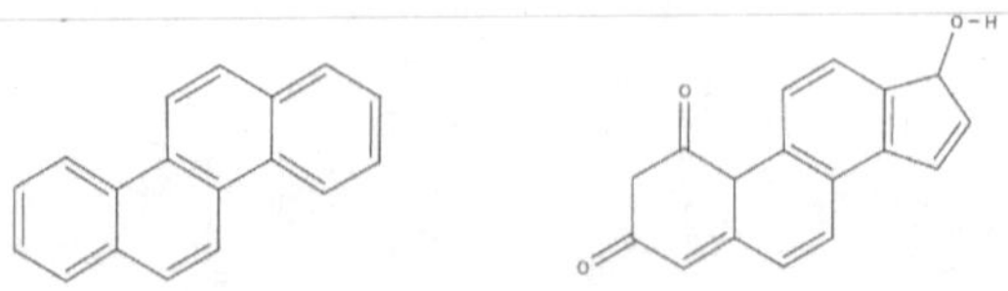

(a) Chrysene (b) Steroid Molecule

Figure 3.1: (a) A starting material of chrysene, which is four fused rings of benzene; (b) the primer molecule derived from chrysene, which is consistent with the analysis on steroid hormones and DNA nucleotides, which I discovered and described in [21, 22], and discussed in Section 2.

reffig:primeRing(b), the rings are oriented left to right and represented by different boxes. So, for example, the four ring structures are represented by individual bars in which the color coding is represented here as orange for the five-carbon ring at the end, cyan for the six-carbon ring which is located towards the back when viewed, green for the six-carbon ring towards the viewer, and blue for the six-carbon ring at the end. The orientation of the ends, whether forward or back, can be determined from an orientation of the inner molecule, as the sequence is back-forward-back-forward, or forward-back-forward-back. This color key will be involved in encoding DNA, and each rectangular box represents one of the rings.

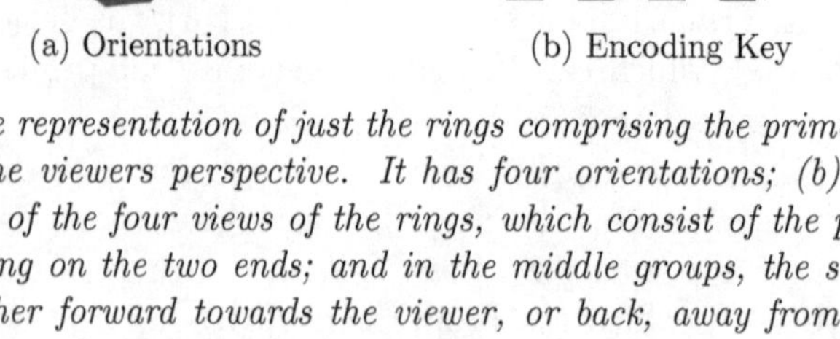

(a) Orientations (b) Encoding Key

Figure 3.2: (a) the representation of just the rings comprising the primer molecule, which from the viewers perspective. It has four orientations; (b) a graphical illustration of the four views of the rings, which consist of the penta-ring and the hexa-ring on the two ends; and in the middle groups, the six-membered rings are either forward towards the viewer, or back, away from the viewer. This can be seen from the ring representation. The encoding key is developed in correspondence with the four orientations of the rings shown in (a). Each boxed rectangle represents a ring, and there is a one-to-one map of the orientation and the encoding key.

3.2 Paired ring structures

The ring structures will now be paired together in association with hydrogen bonding on the ends, oriented as shown in Figure 3.3(a). The alcohol group on the five membered ring will interact with the ketone group of the six-membered ring, and the orientation of the inner rings will associated to match in orientation for aromatic alignment. The corresponding color coded key for the ring structure is shown in Figure 3.3(b). Note that the colors are opposite on the ends, and matched in the middle. These paired molecules are enantiomers, mirror images of each other.

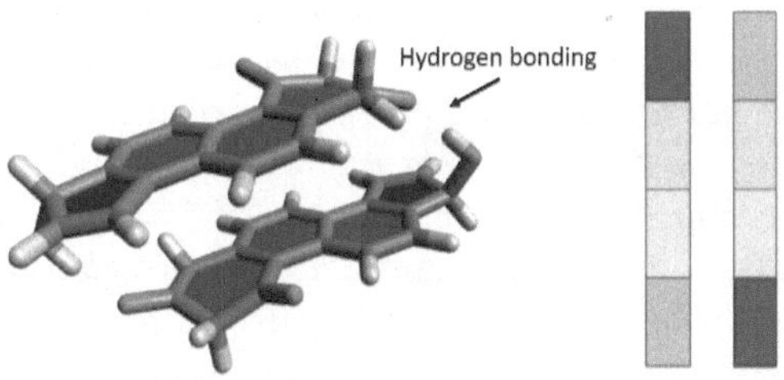

(a) Hydrogen Bonding (b) Paired

Figure 3.3: (a) The developments focus on a paired structure of two molecules comprised of four fused rings, in the shape of a steroid molecule, with three six carbon aromatic rings, and one five carbon aromatic ring, that is functionalized at each end, one by a hydroxyl group and the other by ketones. Hydrogen bonding secures the two molecules. (b) The illustration represents this configuration. This key of using rectangles to represent a ring structure is used throughout this manuscript.

In this paired configuration, there are four orientations possible of the same structure, when viewed from the same perspective, for example left to right. This is illustrated in Figure 3.4, which shows at the left, the sequence top to bottom as, blue, green, cyan, orange in 3.4(a) which according to the represents the penta-ring, forward six-membered ring, back six-membered ring, hexa-ring, which it should be noted has a diketone structure. This configuration can then be rotated and flipped to arrive at the other three configurations. These configurations will be used in creating a unified structure not only for DNA but also paired steroid molecules!

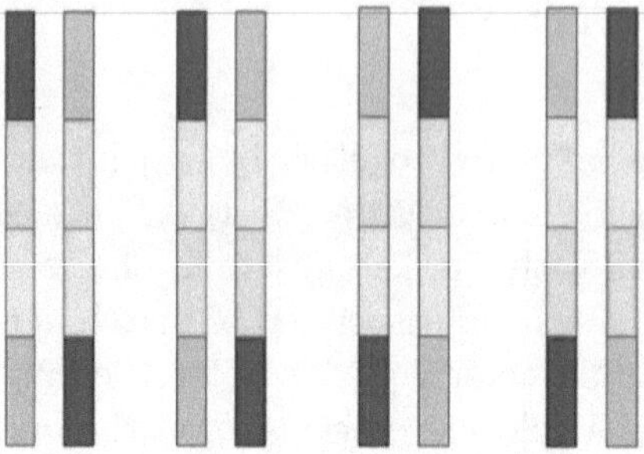

Figure 3.4: The orientation of the adjacent molecule of the second pair with respect to the molecule of the first pair will define the final result. There are four orientations possible. The coupling elements between the adjacent molecules, that is the nature of the intermolecular link, will be used to define the base pairing. Note that there are four rings that define each structure, and that there are four and only four base pairings. Thus starting with (a), the difference with (b) is in the middle even though the ends are the same, which can be obtained by flipping and rotating the molecule. (c) results from (a) by rotation, and (d) results from (a) by flipping. Thus the same pair has four orientations relative to the observer.

3.3 Sequence of pairs

To build-out DNA, the paired molecules of Figure 3.4 will be paired to each other. Thus, for example, the sequence will be established as 3.4(b) followed by 3.4(c) followed by 3.4(b) followed by 3.4(c) and so forth. First though let's take a look at the chemistry of the pairing.

As indicated in Figure A.6, the coordination of the paired building blocks enables a sequence of effective interaction vessels. This confined environment enables the potential for an enzymatic capability in which the temporary formation of bonding arrangements to permit intermolecular cyclic structures that can establish element substitution, the degradation of chemical bonds to reduce intramolecular stress, and the differentiation of the templates into their individual capacity as a nucleotide base. While the structures did have aromatic stabilization, the intermolecular structures can benefit from resonance stabilization, and thus the intermolecular constraints to maintain an in-plane structure during the synthesis steps would have promoted the present-day configuration in which the original aromatic structure was traded for an amine based self-assembled construct of a planar configuration. These chemical concepts are discussed in depth in Appendix A.

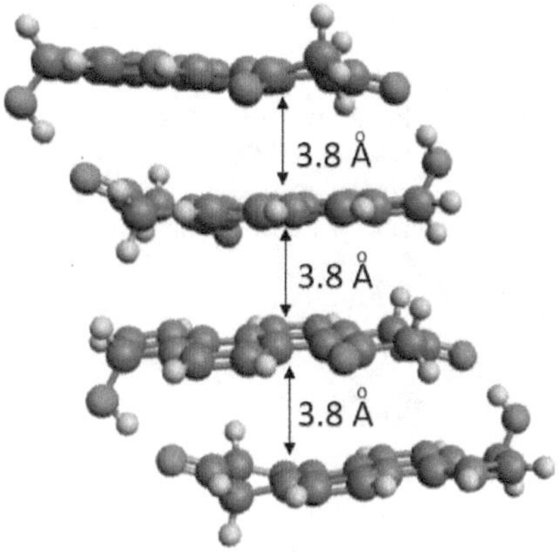

Figure 3.5: There exists the capacity for an co-enzymatic function in the development of each molecule when aligned and stacked. The aromatic nature will stabilize the position, and provide many groups for interaction with the environment. It is indicated that this constitutes a series of reaction vessels, which together with aggressive environmental conditions, the co-synthesis of the DNA nucleotides, and its triggering elements, comprised as steroid molecules, would have resulted from interactive co-development. There are two styles of enzymatic reaction vessels. The directly coupled are used in the synthesis of the coupling to the phosphodiester base, and the vessel that is defined by the meeting of two paired units, is used in the differentiation and separation of the DNA nucleotides and the steroid molecules. It is noted that each of the molecules comprising the units are accessible for the exterior and do not require entrance into the structure.

For encoding DNA and steroid, the main concept is that in stacking the next pair of tetra-ring molecules on top of the initiating pair of tetra-ring molecules, the relative orientation will define ultimately the "letters" of the nucleotide pair, whether it is adenine-thymine, thymine-adenine, guanine-cytosine, or cytosine-guanine. There are exactly four configurations, when placing the second set on top of the first set, and as will be seen later, one molecule of the pair will define the nucleotide pair, and the other will define the cooperative steroid molecule that will interact with DNA to trigger replication and transcription. For example, in Figure 3.6(a), the third molecule starting from the bottom will take on the DNA nucleotide pairing of (in this case adenine-thymine), after the process steps to be described are completed. In Figure 3.6(b), the color code is represented.

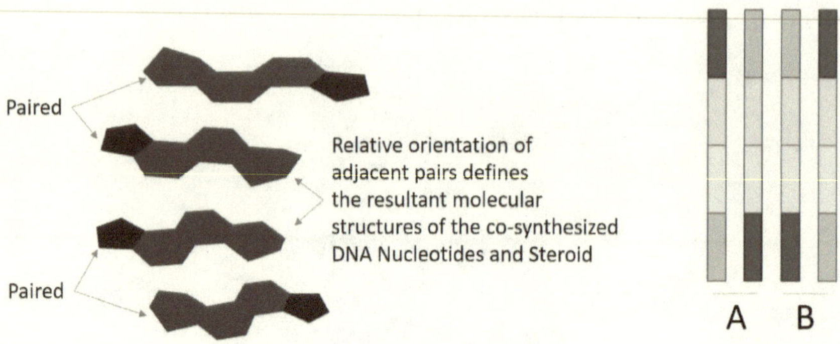

(a) Pair of pairs

(b) Pair of pairs with color key

Figure 3.6: As a major result, the basic code of DNA is formed by stacking two coupled, functionalized steroid molecules, in which here the DNA nucleotide code is defined by the interface of the two molecules orientation. Ultimately it will result in one nucleotide and one steroid molecule. The coding is shown in (b).

Since there are four orientations for each ring structure, the paired structures pairing to each other will result in 16 relative configurations. This is illustrated in Table 3.1, in which a letter code is used for each molecule: P denotes the five-membered end ring in which P stands for penta; B denotes the back orientation of the inner six-membered ring; F denotes the forward orientation of the six-membered ring; K denotes the six-membered ring end molecule in which K stands for its ketone groups. The various orientations are obtained by rotating and flipping, and since there are four possibilities, it will be possible to obtain the next letter independent of the starting point of molecule 1.

Remarkably, that is sufficient to create a coded DNA molecule as well as its steroid molecules, which will enable genetic replication and transcription, that is the results of Figure 3.6. It is possible to create any set of letters from AGCT, as well as its steroid components, by using the strategy of pairing and orientation, which is an amazingly simple process, as it needs to be! Moreover, in addition to the base 4 code associated with the orientation of the pairings, another code is defined: It is also noted that in addition to coding the DNA nucleotides, it also encodes the steroid molecules. As will be developed, the steroid molecules are coordinated with amino acids during the building process, and thus will provide the specificity required for protein translation, as well as an inherent capability to induce strand separation.

Left Pair	Next Pair Options
PBFK	PBFK, PFBK, KFBP, KBFP
PFBK	PBFK, PFBK, KFBP, KBFP
KFBP	PBFK, PFBK, KFBP, KBFP
KBFP	PBFK, PFBK, KFBP, KBFP

Table 3.1: There are 16 possible configurations of pairs of pairs. That is each pair has four orientations, and thus there are sixteen possible arrangements. However, when one set is fixed there are only four possible options.

The steroid triggers have two orientations relative to the upper pair of molecules, meaning that there is alignment in terms of hydrogen bonding to either one side or another. Thus, in addition to encoding what will become the nucleotide pair, it also encodes the type of steroid molecule, as affiliated with its hydrogen bonding location. To further examine the orientation of the base-pairing, Figure 3.4 indicates that there are four configurations, and only four configurations in planar relation to the underlying molecule, which will become its paired steroid.

3.4 Encoding through Intermolecular Binding

When the pairs of molecules are brought into association, there will be an intermolecular interaction in terms of the facing molecules from each pair, herein referred to as hydrogen bonding, implemented by a ketone group on the end hexa-ring of the six-membered end group with the corresponding adjacent molecule of the adjacent pair. As indicated in Figure 3.7, the nature of the interaction will depend upon orientation of the molecules. The internal ring groups may be in phase or out of phase. It will turn out that this phasing will define whether the paired set of molecules becomes from the set A-G-C-T and will also determine the structure of its corresponding steroid. The orientation defines whether the intermolecular coupling is achieved with the oxygen group on the same side or on the opposite side.

In order for intermolecular hydrogen bonding, or interaction of oxygen functional groups, to occur between pairs of pairs, there has to be the ketone present, and there has to be potential for it to couple to a second group. If it is in-phase, that is the oxygen groups on the same side, then interaction can occur, otherwise it will not happen. This can be expressed in color coded comparison of the 16 possible structures as indicated in Figure 3.8, which indicates that of the 16 possible configurations, 8 will have hydrogen bonding capability, and the other 8 configurations will not have

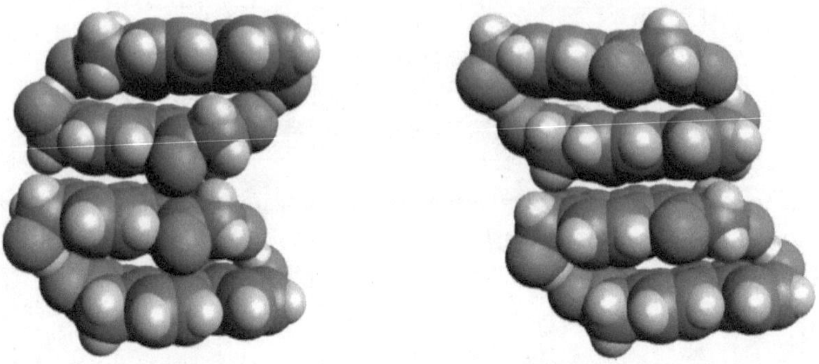

(a) O on the same side of coupling (b) O on opposite side of coupling

Figure 3.7: (a) When the rings are oriented into two of the four possible configurations, the oxygen molecule that is available for intermolecular binding are on the same side. This will involve interaction, but and ultimately hydrogen bonding, and thus a differentiation into the co-synthesis of the steroid molecule and DNA nucleotides, that result. (b) On the other two configurations, the oxygen molecules are not on the same side, and thus will result in a different synthesis sequence, and different nucleotides, and will preclude inter-pairedhydrogen bonding. Likewise, the orientation of the end groups will provide a second independent binary code, and thus the combination of two binary codes will enable a base 4 code.

hydrogen bonding capability.

In addition, for each configuration, there corresponds 2 configurations that will have intermolecular hydrogen bonding capability, and 2 configurations that will not have intermolecular hydrogen bonding capability. This is illustrated in Table 3.2. From the sequences, it can be seen that if the next molecule starts with the same molecule, then the correspondence has to perfectly matched in order for hydrogen bonding to occur; or if the next molecule is not the same, then it has to be completely out of phase in order for hydrogen bonding to occur.

3.5 Summary on Encoding DNA

The encoding of a four-ring molecule (which will become DNA nucleotides co-synthesized with steroid molecules that enable function, as will be shown) as an independent base 4 sequence has been indicted. Moreover, a second set of four-ring molecules (which will become matched steroid molecules) has also been indicated, linked by hydrogen

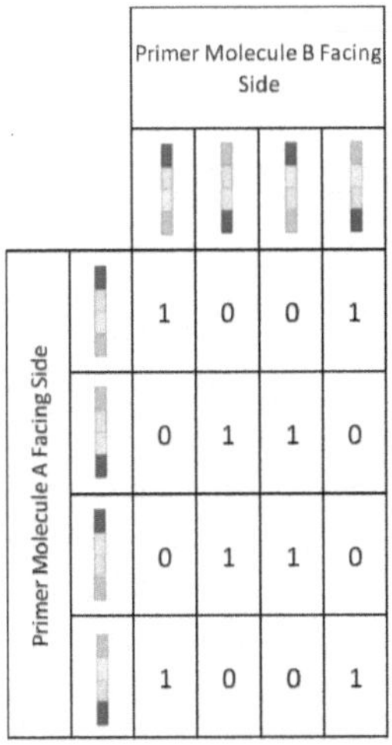

Figure 3.8: The intermolecular orientations that will lead to hydrogen bonding. Out of the 16 total configurations, eight will lead to intermolecular hydrogen bonding. These couplings enable a base 4 code for the nucleotides, as well as a base 2 code for the associated steroid molecules.

bonding has also been shown, which will comprise a set of base-two molecules because of its dependence. Both sets are needed to achieve structure and function. Long sets can be strung together. It can also be represented as a long sequence of random molecules, as indicated in Figure 3.9. This is established by selecting one of the four from Table to determine if there is hydrogen bonding or not. This will result in a long DNA chain, and a set of steroid molecules, which will thus enable Structure and Function, respectively. It will be seen that the steroids are the workhorse of the implementation of DNA.

Thus, by using this mechanism of assembled paired representations in configurations, long sequences can be arranged in which every other pair will have a random orientation drawn from a set of four possibilities. For example, the sequence of paired molecules is indicated in 3.9 which has a first base 4 code set at molecules 1 - 3 - 5 - 7 . . ., and a second base 4 code set at molecules 2 - 4 - 6 - 8 . . .. However, the first base 4 code set is correlated to the second base 4 code set because they were presented as a pair. Thus, the result will be one independent base 4 code set, and a second dependent set that can be derived from the first set. We will see that the first set corresponds to A - G - C -T DNA, and the second set will be the steroid molecules, pursuant to the discovery discussion in Chapter 2 on the correlation of

Referencing Pair	Next Pair Intermolecular Bonding	Next Pair No Intermolecular Bonding
PFBK	PFBK, KBFP	KFBP, PBFK
KFBP	PBFK, KFBP	KBFP, PFBK
PBFK	KFBP, PBFK	PFBK, KBFP
KBFP	KBFP, PFBK	KFBP, PBFK

Table 3.2: This is an important table. It indicates the orientations that will lead to intermolecular hydrogen bonding when the pairs are stacked together. The intermolecular bonding will define whether the resultant nucleotide pairing is A-T or C-G. Note that intermolecular hydrogen bonding means adjacent steroid molecules. Together with the orientation of the first group, it will define the base 4 code. It will also introduce a coupled code through the related steroid molecule, which will have two structural classes, with and without a mid-section, intramolecular, oxygen based functional group.

DNA nucleotides and steroid molecules.

Figure 3.9: This is an example of an encoded molecular structure that ultimately will form the DNA double helix and the companion steroid molecules. The steroid molecules enable opening of the DNA double helix to allow access to the strands. It also comprises walls of a reaction vessel, and also as a transport mechanism to move the amino acids into the interaction vessel.

Chapter 4

Transmitting DNA Structure and Function

In order to transmit the information enabled by the structural representation of the DNA molecule, it has to be packaged in a stable configuration. Moreover, it has to be accessible such that it can be decoded and used at the receiving end. In this chapter, the procedure to package DNA and its machinery necessary for its function is determined. The result is the DNA assembled as a double helix molecule, along with triggering agents, that is steroid molecules, which permit it to be accessed, decoded, and translated, as well as error corrected. In this section, to present the transmission of the encoded DNA complex, the following outline is applied:

- Description of the enzymatic complex that ultimately forms into an interleaved structure of complementary base paired DNA nucleotides matched to a steroid molecule of one of two types.
- Synthesis of the final products arranged still in the enzymatic complex format: these steps are largely described in Appendix A.
- Release of the final products to their final form of the DNA double helix, comprised of four complimentary base pairs, and the steroid molecules that are necessary for decoding the DNA double helix, translating the encoded structures into a string of amino acids that form proteins.

4.1 Co-enzymatic unified complex

The molecular structures comprise co-catalytic environment that involves the interaction of one molecule on the other, such that one influences the shape of the other, and thus a co-formation process. An example is depicted in Figure 4.1(a), the stack of ten molecules for the creation of the five nucleotide pair of (T-A), (A-T), (G-C), (T-A), (C-G), starting from the bottom, is presented. The other five molecules will be associated with steroid molecules, which can be used for replication and transcription. Both will be developed in the same complex, and both are interactive and influence each other. The nucleotides will form on molecules 1, 3, 5, 7 and 9, and the steroids will be formed on molecules 2, 4, 6, and 8, with the numbering starting from the bottom. There is another molecule below molecule 1 that influences its configuration, and would ultimately result in a steroid, but that molecule is paired, and in order to keep it brief, only the transformation on molecule 1 will be shown, and not the corresponding change to its paired molecule, which would be situated below it. Further, the starting orientation of molecule 1 is taken to indicate the independence of the letter sequencing. The top molecule 10 only couples to molecule 9, and does not influence its configuration.

4.2 Synthesis of unified complex

Following the suggested reaction mechanism outlined in Appendix A, an initial step is represented in Figure 4.2 of linking the phosphodiester backbone, to the sequences, which is shown for the repeating AGCT unit, here shown as if T preceded A, and A followed T. The phosphodiester is linked through covalent matching to the sugar arrangement to molecules 1, 3, 5, 7, ... and through hydrogen bonding to molecules 2, 4, 6, 8,.., of which the suggested reaction mechanisms indicate a perfect size arrangement with regard to the spacing of the DNA molecule.

The next step in the sequence involves the application of the intermolecular enzymatic complex to introduce differentiation and separation into the covalently linked molecules, thereby forming the DNA nucleotides. This is illustrated in Figure 4.3, in which the orientation of the molecules is noted, and the nucleotides are noted. The steroid molecules are also indicated by the color coding. The steps necessary to produce this complex is described in the Appendix. Here we are just interested in the coding principles and transmission schemes, and thus the interesting details of the chemical reactions are not described.

4.3 Simultaneous release to form the double helix and steroid molecules

Even though the DNA nucleotides have been formed, it is not in a format that is stable for transmission, and thus a method is needed to stabilize the unit. And in

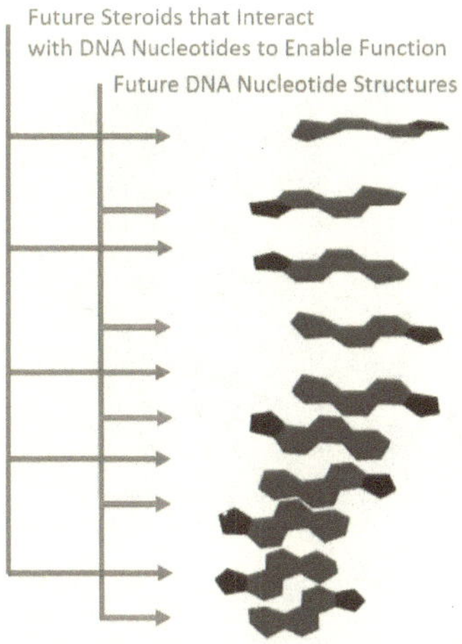

(a) Ring Representation

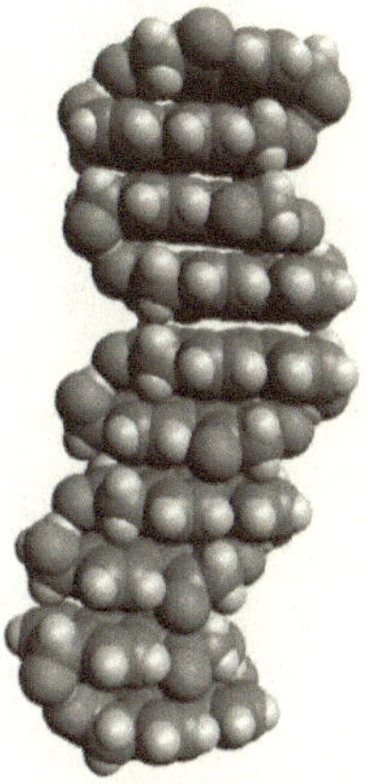

(b) Co-enzymatic complex

Figure 4.1: (a) A co-enzymatic structure resulting in the DNA nucleotides and steroids that interact with the DNA nucleotides to enable function after synthesis. It will also include four steroids configured as corticosteroids similar to cortisol, for molecules 2 and 6, and similar to testosterone from molecules 4 and 8. The tenth molecule would interact with the eleventh molecule. (b) The van der Waals spheres indicate hydrogen bonding along two edges. Note the code along the middle of the middle of the molecules represented as ketone molecules, in this case from the bottom 0110100101. The other independent base 2 code comes from the edge of the molecule, which enables a base 4 structure.

the process of stabilizing the unit, there is the amazing situation of putting forth in motion, structure and function, and thereby instantaneously creating life function!

In the process of pairing and pairing, it is noted that some of the nucleotides are out of phase with respect to the nominal orientation, as shown in the illustrations of Chapter 2. Thus, there needs to be an orientation step of rotating these molecules, which are out of phase, to in-phase such that a stable structure can be formed. This structure is the double helix, and in the process it will release the steroid molecules during the rotation of the nucleotides.

It is seen in the illustrations of the double helix that the purine molecules are in a forward position, while the pyrimidine molecules are in a back position, and this configuration needs to be achieved to stabilize the molecule. On average, if it is a random sequence of letters, half will be in phase and half will be out of phase. The

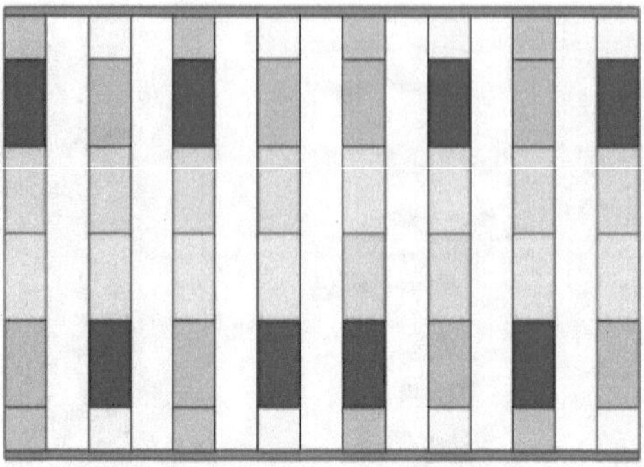

Figure 4.2: Binding a phosphodiester backbone and connection to the DNA nucleotides while securing the accompanying steroid molecules through hydrogen bonding achieves a unified complex that enables co-enzymatic activity in forming the DNA nucleotides and completing the overall procedure of achieving both structure and function for information storage and reading.

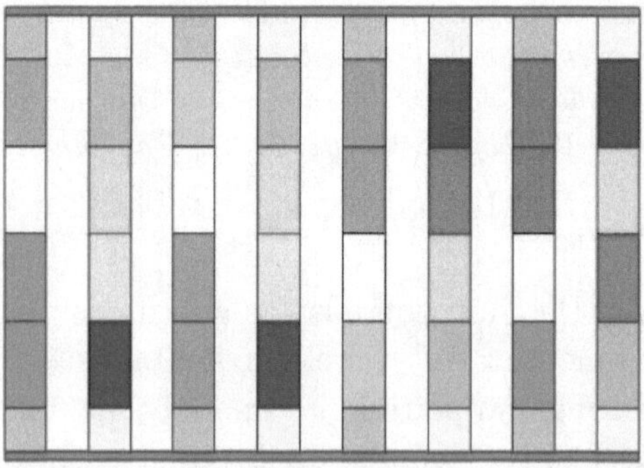

Figure 4.3: The synthesis steps result in the DNA nucleotides separated into complementary base pairings, coupled through hydrogen bonding through base pairs, as well as through intermolecular hydrogen bonding.

orientation with all molecules in phase to their nominal position, and the release of the steroid molecules is indicated in Figure 4.4. Here shown for the molecules AGCT. This structure is stable as we know by the double helix format.

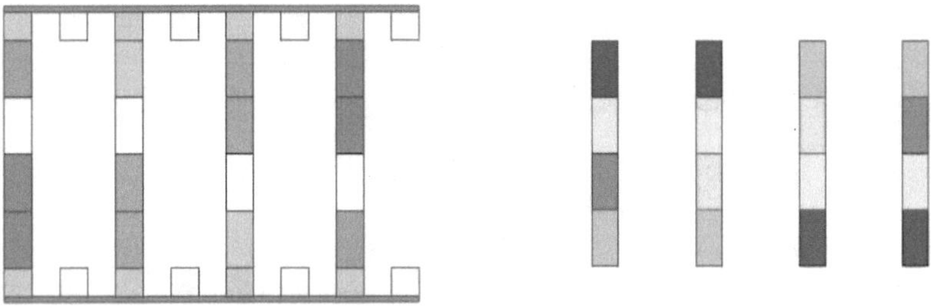

(a) DNA Double Helix (b) Released Steroids

Figure 4.4: (a) Releasing through the formation of the double helix, which involves a re-orientation of the purine and pyrimidine molecules such that there orientation as seen along with one strand is opposite. That is the pyrimidine are facing forward relative to the second ring, whereas the opposite is the case for the purine molecules. (b) The release of the steroid molecules through the rotation necessary to achieve the double helix indicates that there will be equal number of steroid molecules to that of the co-synthesized DNA nucleotides. There are two types of steroid hormones produced, thus a base 2 construct.

In the ejection of the steroid molecules, two types are formed, as shown in the Appendix. These are represented in Figure 4.5 of the planar molecules which have two types, one with an intramolecular hydroxyl group, and the other without one. These molecules are the precursors to steroid hormones, but for now, are used for other functional purposes.

Note that the flipping does not take place TA zones and CG zones. Thus, when the double helix formed, there were some areas that were already in the double helix shape and did not require flipping, thus the steroid molecules remained intact. This provided a zone of stability, but the inability to move, and thus created small zones of stability about it, which resulted in the separation of the strands. Maintaining the integrity of these regions. Immediately the strands were thus accessible and capable of generating amino acids.

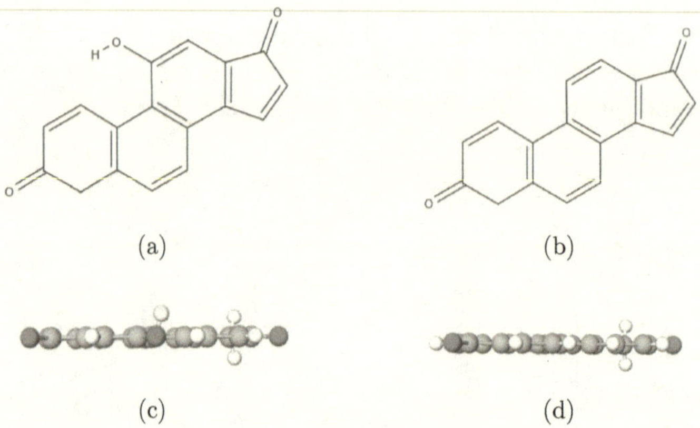

Figure 4.5: In addition to the DNA nucleotides, steroid molecules are produced by the unified complex. The steroid molecules have a major role to play in enabling life function for DNA, in addition to providing catalytic and differentiating capability to produce DNA in the first place. There are two types of steroid molecules ultimately produced, and later released during the transition to the double helix are: (a) with a hydroxy group in the mid-molecule and the other (b) without a mid-molecule hydroxy group. In (c) and (d), it is noted that both molecules are planar constructs and thus are consistent with the structure of the DNA nucleotides.

4.4 Strand separation to access DNA

During the flipping process, some regions that are highly regular with alternating regular regions of TA and CG will not require flipping since they are already in position. Thus, these areas will stabilize the strands and there will be an opening.

While the DNA nucleotides are now in a stable double helix configuration, in order for it to have purpose, it needs to be accessible for reading and action. It is good that the double helix is so stable, but it also limits its availability for opening and reading, and without that capability it is just an interesting polymeric molecule, but would serve little purpose. Fortunately, it was co-synthesized with its key, the steroid molecule, that is capable of opening it. Therefore, in a transmission format, it is necessary to transmit not only the DNA molecule but also its key to opening it. Therefore, it is favorable for DNA, and its paired molecules, to be coupled in a cell format to contain the molecules. It is possible to deploy multiple steroid molecules to induce a stronger association and effective binding of the DNA strands, and thereby improve the duration of the subsequent opening of the strands of the nucleic acid.

Thus, with intermolecular binding the strongest, it would favor chains of TA to form stabilization sites (TA chains), or longer chains of the weaker CG interactions (CG chains). Strand separation is described in the Appendix through mathematical modeling.

Thus, having produced the paired and encoded structures of both DNA and steroid molecules, the bases will only be accessible for replication and transcription or translation if the strands are possible to separate. Fortunately, the steroid molecule is available, and can bind onto the phosphodiester chain to enable strand separation. This occurs through stabilization of the strands, which are vibrating and oscillating molecular constructs. Thus there is kinetic energy transferred to potential energy through the stabilization process, which is an overall energy function whose internal energy must be dissipated through for example molecular disorganization, and bond breaking in this case. As the DNA molecule is stabilized by binding with the steroid molecule, there must be a balance in energy in terms of a release through bond breaking to be nearly equivalent to that of the stabilization provided by the binding event. Essentially, the DNA strands act as two strings in constant motion, which is fluidic, and with the cross-bonding induced by the steroid molecule, that is the triggering element, it acts as a pinch-point, and the strands, which were in relatively independent motion become constrained at one point, and around that point, a small bubble must form through hydrogen bond breaking upstream and downstream of the pinch-point. In Figure 4.6 the result of an open single-strand DNA.

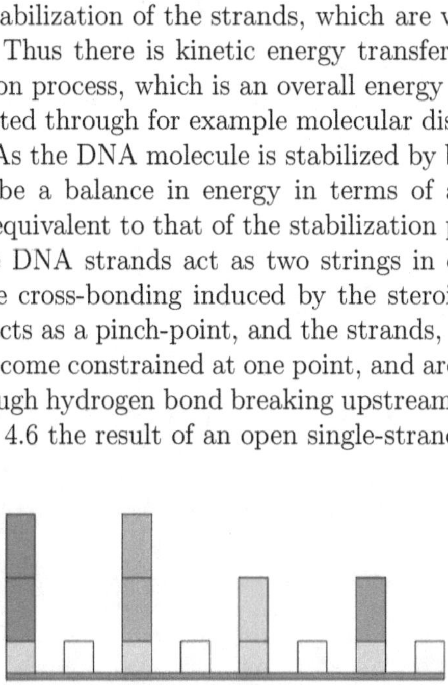

Figure 4.6: Double helix opening through the binding of steroid molecules at initiation sites, thereby separates the double strands and completes the transmission of the information process, as the message is received, therefore making available the DNA nucleotides for error analysis and decoding. The single strand of AGCT is represented here, connected through a single phosphodiester backbone and sugar.

After encoding DNA as comprised of a set of chemical molecules, their transmission as a sequence of DNA nucleotides arranged as a double helix, which was accomplished in conjunction with steroid molecules as part of a unified complex that co-catalytic, with one molecule influencing the other to arrive at a final product set. Both are useful as the DNA molecule provides the library, and the steroid molecule provides access to it, by separating the strands through a binding operation. The dramatic nature of the rotation of the unified complex into the double helix thereby creates an instantaneous moment of the convergence of structure and function, and thereby a life function of a genetic purpose.

Chapter 5

Decoding DNA

Having formed DNA, its companion steroid molecules, and having the ability to access its code through the binding of the steroid molecules to the DNA strands, it is now necessary to decode it. In this chapter, the steps necessary to decode DNA are described. To differentiate this step from translation, I will take the decoding step up until the point where its interaction with amino acids occurs. With this chemical procedure, the prebiotic production of encoded proteins originating from nucleic acids can be precisely defined. The steroid molecule is a key component, as it enables the guidance of the DNA molecule, thereby encoding both itself and the DNA structure, and also effectively results in encoded amino acids, which can be mapped to the DNA sequence through the intermediary steroid molecule.

5.1 Nucleotide error checking

In the context of initiating the decoding process after receiving the message, the steroid molecules align to each nucleotide, and thus in effect test the integrity of each nucleotide. This would involve the capability to mirror the nucleotide, including its presence. This is indicated in Figure 5.1, in which each steroid molecule is coupled with a corresponding nucleotide, depending upon whether hydrogen bonding is capable. Thus, the steroid with a hydroxy group will pair with A-T and those without one will pair with C-G. In the Appendix, error correction is discussed in which it is pointed out that the original orientation and code can be determined precisely by knowing the starting point orientation. Note also that by error checking, it also has the benefit of establishing a structure with a uniform point. As indicated in Figure 5.1, the strand opening provides access to the steroid molecules, and leads to an association with its steroid molecule. This forms a sequence of alternating steroid molecules in the same height, with shorter nucleotide bases in between each uniform structured steroid molecule.

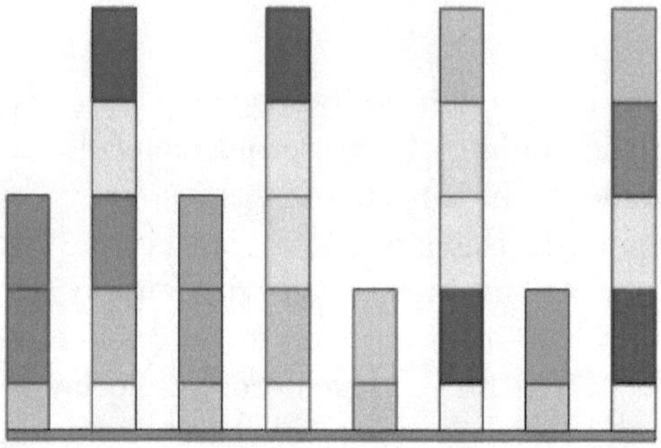

Figure 5.1: As an initial step after the message is received as transmitted, in starting the decoding process, error checking is performed on the bits by pairing the steroid molecules, which also must be in the vicinity of the DNA nucleotides, in an aligned manner. The appendix includes additional discussion on error correction codes, which are possible due to the PBFK process described in its encoding. The result is a structure with steroid molecules alternative the DNA nucleotides.

As a basic result shown in Figure 5.11, the opening of the strand and association with steroid molecules through hydrogen bonding, enables a repeating self-contained interaction vessel as indicated by the dotted line. Within each set of three, there will be a zone which is comprised of two walls: one wall is the steroid molecule affiliated with nucleotide one, and the second wall is affiliated with nucleotide two. Moreover, each wall is supported by the nucleotide on the opposite side of the dashed enclosure, and thus the third nucleotide is needed to achieve structural symmetry of the zone. Furthermore, the third wall, while it does not significantly contribute to the direct formation of the interaction vessel, it will serve as an alignment and docking post.

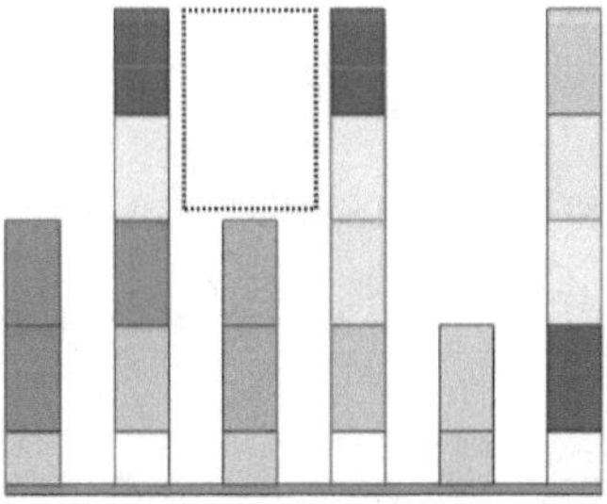

Figure 5.2: As a basic result, the opening of the strand and association with steroid molecules through hydrogen bonding, enables a repeating self-contained reaction vessel as indicated by the dotted line. Within each set of three, there will be a zone which is comprised of two walls: one wall is the steroid molecule affiliated with nucleotide one, and the second wall is affiliated with the third wall. Moreover, each wall is supported by the nucleotide on the opposite side of the dashed enclosure, and thus the third nucleotide is needed to achieve structural symmetry of the zone and will later serve as an alignment and docking post for delivery of amino acids to the interaction vessel zone.

5.2 Decoding table for the interaction vessel

Referencing the procedures described for encoding and transmitting DNA as a double helix formed as an intermolecular enzymatic complex with steroid molecules, in Figure 5.1, which is a depiction of the pairing of the Adenine nucleotide with its steroid molecule, the Guanine nucleotide with its steroid molecule, the Cytosine nucleotide with its steroid, and the Thymine nucleotide with its steroid, the chemical decoding key is enabled, which is needed to determine the molecular interaction with the nucleotide and its matched hormone. This will be the view when the strands of the double helix are separated, and the steroid molecules are matched up to the corresponding base, such that is structural and chemical compatible.

By combining the reference key of Figure 5.1 with all possible configurations of the triplet of nucleotide-steroid pairings according to the results of Figure 5.11, comprising the interaction vessel, it is seen that there are 64 possible configurations which can be arranged in table format of individual tables for Adenine in Figure 5.3, for Guanine in Figure 5.4, for Cytosine in Figure 5.6, and Thymine in Figure 5.5. Thus, a particular triplet can be picked, and the structural and chemical nature of its associated reaction vessel can be determined. Because each reaction vessel is different, depending upon the sensitivity of the chemicals, it is possible to arrange for 64 different molecular interactions. In Chapter 6 on translation, I will show how each reaction vessel has a structural configuration that corresponds to the 20 standard amino acids. For the remainder of this chapter though, I will show how the protein is formed.

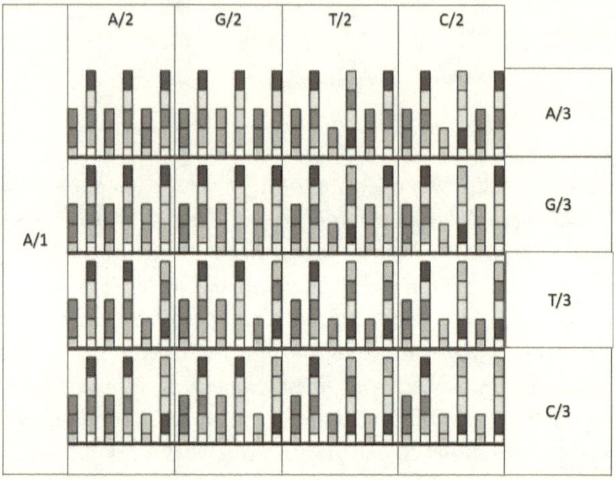

Figure 5.3: In defining a unique interaction site for each set of three coupled nucleotide-steroid pairings, the different reaction zones are coded for the nucleotide starting with adenine. Note that results a set of sixteen unique configurations that start with adenine. Thus the first wall is the same for each unit, but differs in the construction of the second and third sites.

5.3 Association of Steroid to amino acids

To stabilize the structure, a supporting molecule across the top of the steroid molecules would be helpful. This would retain the steroid molecules. Fortunately, the steroid molecule itself is a useful structure, because it is planar, and further will be available since it travels with the DNA molecule. Moreover it contains two functional ends, which will permit multiple levels of interaction in structural construction and material transport.

Importantly, the steroid molecule can also interact with either end of the amino acid, as indicated in Figure 5.7. This transport vehicle can then be used to position an amino acid within the vessel and connect up to the next amino acid. Its size is

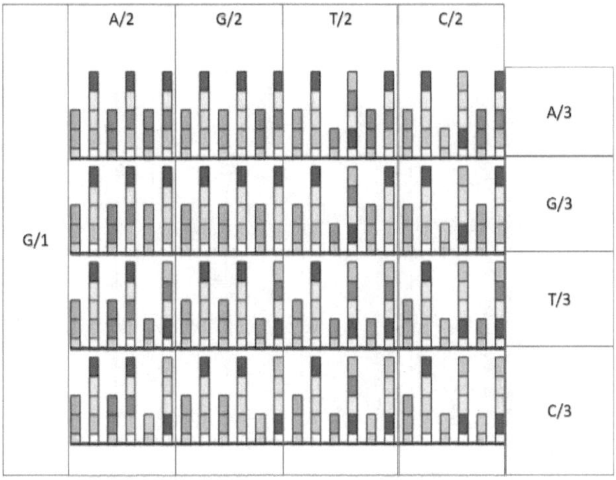

Figure 5.4: Similar to adenine, a unique set of interaction zones is achieved that start with guanine. Like adenine, guanine is purine molecule, and there are unique interaction sites of sixteen associated with guanine.

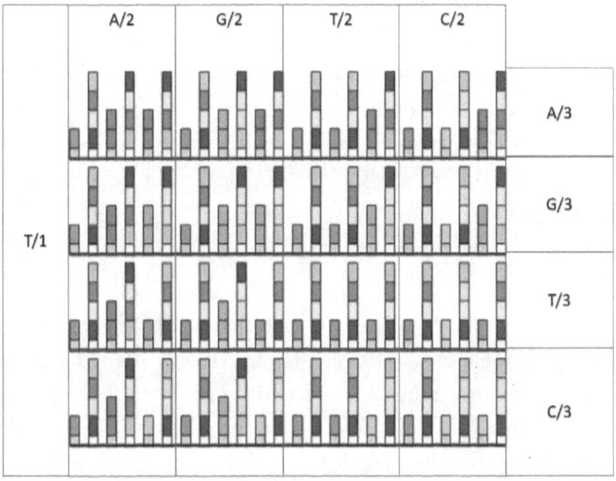

Figure 5.5: Another sixteen sites are unique for those that start with thymine, a pyrimidine molecule.

perfectly suited for the interaction, and has the advantage of blocking off the non-middle reaction vessels. By associating it within a single direction, it further filters the amino acids, to just the L-amino acids, in order the residues to project down the

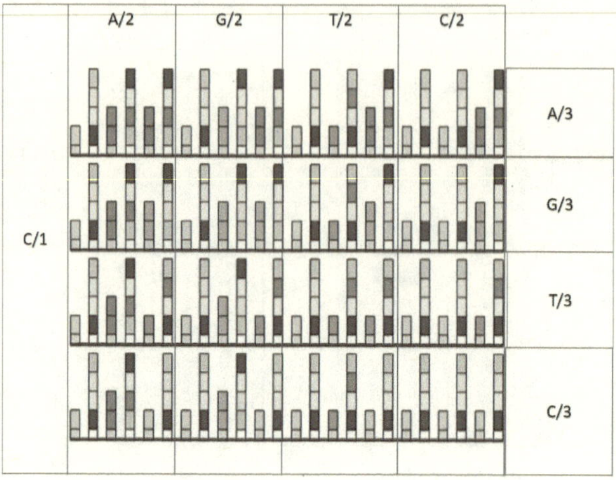

Figure 5.6: Completing the set of interaction vessels is cytosine. Thus, combined with adenine, guanine, and thymine, there are a total of sixty-four unique reaction vessels.

reaction vessel.

To initiate decoding, the steroid molecule, which is the workhorse of DNA functionality, again is used by association with the amino acid. The hydrogen bonding of one end is attracted to the amino group. to bring the amino acid into association with the complex to utilize the strands after checking the integrity of the nucleotides, which also has the benefit of positioning a steroid molecule, amino acids are associated with the strand. After the strands are separated, the coupling of steroid molecules are performed.

This association of the steroid holder is further amplified in Figure 5.8. It indicates that in Figure 5.8(a) that there is a hydroxyl group at the five-membered ring as opposed to the ketone group of the wall structure, other than that all things remain the same. In Figure 5.8(b), the planarity of the molecule is indicated, which will make a strong stable structure for interaction as a ceiling to the wall groups. In Figure 5.8(c), its attachment points are indicated.

5.4 Vessel for amino acid assembly

In examining the structure, it is noted that three consecutive nucleotides form a supported reaction vessel, with respect to the middle group, as indicated in Figure

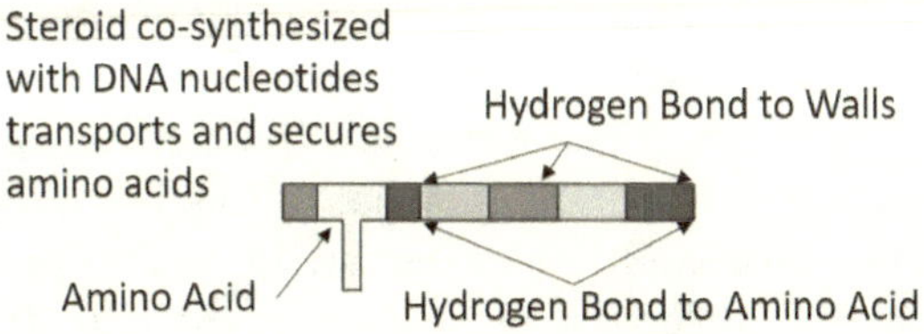

Figure 5.7: Further serving as the workhorse of DNA function, the steroid molecule not only provides the walls of the interaction zone, it also transports the amino acids for association with DNA. The coupling of the amine group to the ketone group of the steroid molecule, which was released through the co-enzymatic formation with the DNA nucleotides.

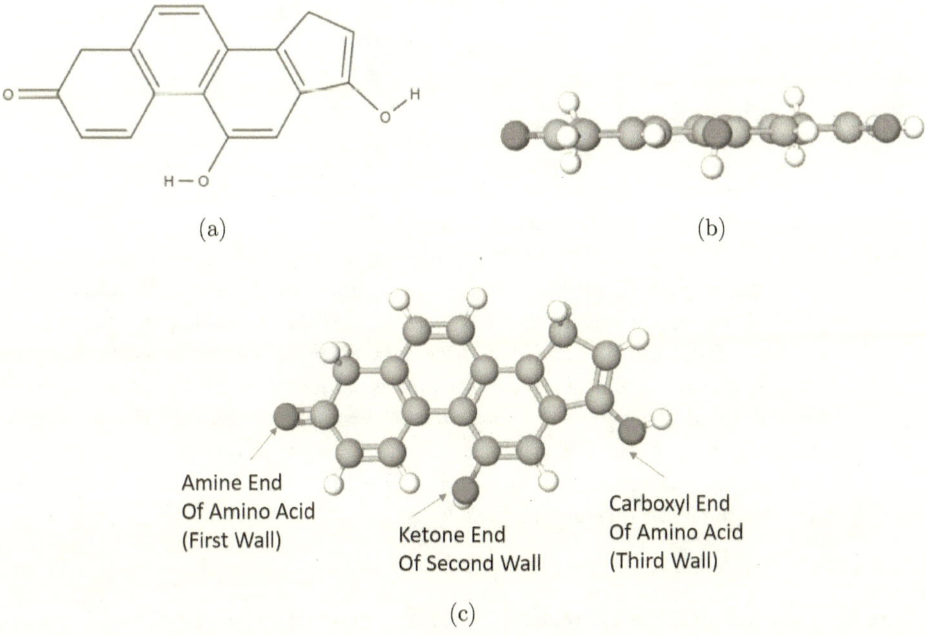

Figure 5.8: (a) The chemical structure of the molecule comprising the ceiling and amino acid holder. (b) The side view indicates planarity. (c) The holder contains three functional groups to link to the three walls comprising the interaction vessel.

5.9. There are two walls and a base. The two walls are supported by neighboring nucleotides on either side of the left and right wall. There is a third steroid molecule that does not form a wall, and is there because the third nucleotide is supporting the right wall, in a manner corresponding to that of the first nucleotide. In addition, this vessel has aromatic walls, and has the possibility for a functional group in the oxygen group of the steroid molecule. Thus, there is the potential for an interesting interaction with outside molecules, and as we will see, this molecule is the amino acid.

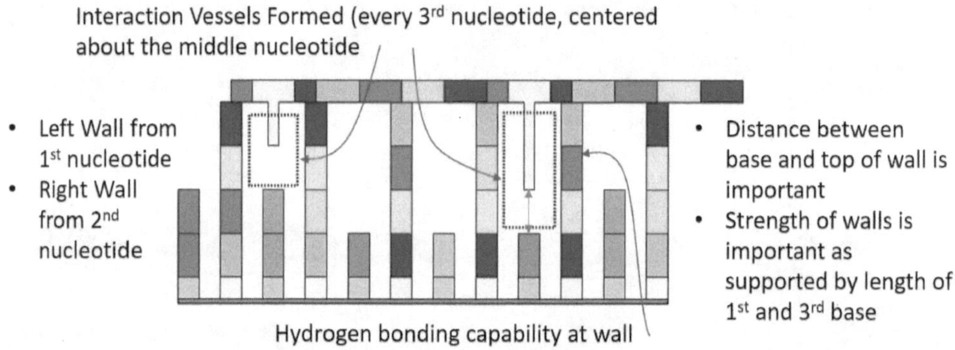

Figure 5.9: The coupling of the amino acid-steroid complex to the interaction vessel forms an enclosure within which the DNA nucleotide will influence the structure and chemistry type of amino acid that can associate it. There are multiple interactions of structure and chemical associated, which includes the distance between the base and the tip of the entrance, the chemical activity of the walls, the strength of the walls, the interaction with the base. Thus, there is a high degree of sensitivity which will permit differentiation of amino acid depending upon the nature of the interaction vessel.

Note that a form of error correction could also take place, if the steroid molecule can not planarize properly it will enable time for a correction to be made to the error in question, such as the replacement of the proper steroid molecule, or if possible the nucleotide base could be repaired or neutralized so that further interaction could take place. The size of the interaction vessel and its opening would also provide a strong filtering mechanism to limit the type of materials entering into the unit, or if one is in place, its disassociation in favor of the steroid - amino acid complex that could stabilize the vessel.

5.5 Structural and chemical nature of the assembly vessel

The structural and chemical characteristics of the vessel defined by the three nucleotides is further studied. In Figure 5.9, the structure of the interaction vessel over the second nucleotide has a major and highly sensitive influence over the type of amino acid that is incorporated within the reaction vessel, which is to be expected. The biggest factor is the distance between the top of the steroid molecule, which is the entry point for the residue of the amino acid and the top of the base. For the purine bases the distance will be greater than the pyrimidine bases. Other factors include the position of the oxygen group on the wall; the support provided by the adjacent nucleotides for the first and the third also play a significant role. The pyrimidine groups will provide less support than the purine molecules because of its height. On the other hand, entry into the vessel will be more hampered by the purine groups because the wall will not be able to move as much, therefore more stability of the wall .

The structure of the interaction vessel over the second nucleotide has a major and highly sensitive influence over the type of amino acid that is incorporated within the reaction vessel, which is to be expected. The biggest factor is the distance between the top of the steroid molecule, which is the entry point for the residue of the amino acid and the top of the base. For the purine bases the distance will be greater than the pyrimidine bases. Other factors include the position of the oxygen group on the wall; the support provided by the adjacent nucleotides for the first and the third also play a significant role. The pyrimidine groups will provide less support than the purine molecules because of its height. On the other hand, entry into the vessel will be more hampered by the purine groups because the wall will not be able to move as much, therefore more stability of the wall.

The chemical aspects are outlined in Figure 5.9. The chemical interactions, like structural, have a significant role on the type of the amino acid that can enter the vessel. Of primary importance is the chemical interactions with the base, if there is size compatibility, as well as the interaction with the functional groups on the walls of the vessels. The chemical interactions with the first and third nucleotides are far less important, but can play a role as it may pull the nucleotide away from the wall, and thus impact the structural support. Thus, the interaction vessel provides a differentiated interaction zone for chemical association.

5.6 Interaction of Vessel and Amino Acids

This interaction is shown in Figure 5.10. Here the candidate amino acid can be brought into contact with the interaction vessel. If stability can be achieved through interaction of the candidate amino acid with the walls comprising the interaction vessel, and if binding is significant, then the unit comprised of amino acid and transfer steroid can remain until removed. If it is not compatible, then the amino acid-steroid assembly will move along to the side, and another candidate will try to lock onto the interaction vessel.

The interaction of the steroid-amino acid can form chains since the two ends of the steroid molecule are functionalized, and both ends of the amino acid are also functionalized. The steroid will cover the nucleotides that are not engaged and thus will make best use of the amino acid material. Furthermore, the steroid on top of the structure will stabilize the structure and enable the interaction of the amino acid with the walls and the base nucleotide to determine if a match is met. This can be illustrated in the example shown in Figure 5.10. An example of how it is put together for three amino acids, which require three reaction vessels and nine nucleotides, since each reaction vessel is located on the middle nucleotide of the triplet, in which the first and third nucleotides impact the structural and chemical characteristics of the interaction vessel. The connector blocks the third and first openings while the interactions are taking place in the second reaction vessel. The size of the connector which is a steroid molecule fits precisely the configuration. When the interaction is stabilized, it rotates out of the vessel for attachment with the next amino acid in forming the protein

5.7 Assembly of the decoded amino acid chain

Now having formed a stabilized structure, it is necessary to form the protein. This can happen if polymerization ceases, for example by hitting a site that does not permit further stabilization. In Figure 7.7, a mechanism is shown in which the connector detracts from the molecular structure, which allows the amino acid to be directed in close proximity to the next amino acid, and achieve a peptide bond. This will continue until the end is reached. While the detachment of amino acid occurs, the corresponding steroid is free to leave and find another amino acid for attachment and to search when it can be applied.

Another example is presented in Figure 5.12 for five amino acids are coupled into the organized structure in which steroid molecules form the walls and ceiling of the reaction vessel, all contained through hydrogen bonding including the steroid on the

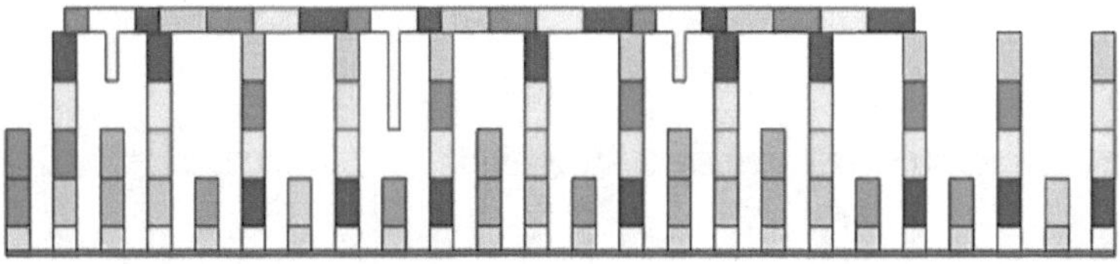

Figure 5.10: An example of how it is put together for three amino acids, which require three reaction vessels and nine nucleotides, since each reaction vessel is located on the middle nucleotide of the triplet, in which the first and third nucleotides impact the structural and chemical characteristics of the interaction vessel. The connector blocks the third and first openings while the interactions are taking place in the second reaction vessel. The size of the connector which is a steroid molecule fits precisely the configuration.

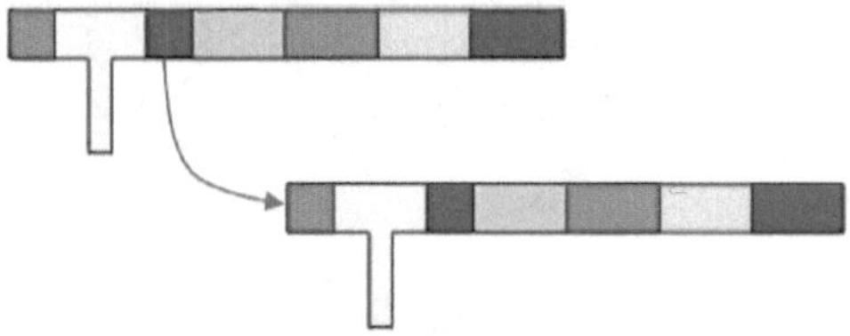

Figure 5.11: To produce a protein, after there is a period of stabilization dependent upon whether there is an adequate fit, the workhorse of DNA function again enables the process. There is a removal of the amino acid-steroid complex driven by the coupled nature of the interaction with the chain of amino acid-steroid, and will enable one amino acid to interact with the adjacent amino acid to form the peptide bond, and a protein.

celling to each wall. The stability and chemical interactions are defined through the triplet nucleotides, and thus provide a differentiated reaction environment for repeated assembly of amino acids.

In Figure 5.13, the result is indicated. After the amino acids are coupled through rotation out of the complex aided by the steroid molecule, a protein results whose characteristics are defined by the vessel from which it demonstrated stability. Thus, the process is completed and repeats as long as the strands remain open.

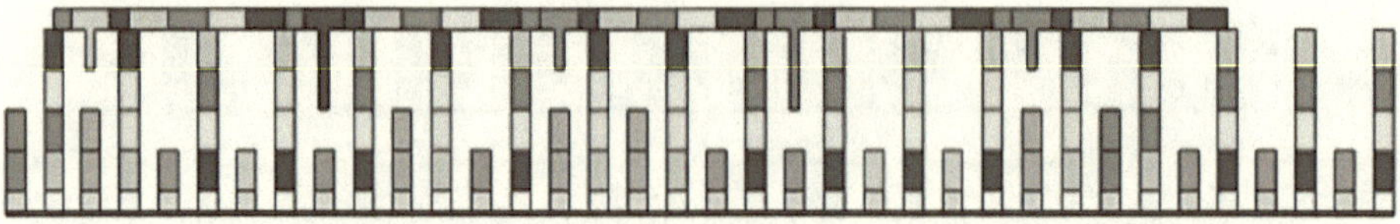

Figure 5.12: As an example, five amino acids are coupled into the organized structure in which steroid molecules form the walls and ceiling of the reaction vessel, all constrained through hydrogen bonding including the steroid on the celling to each wall. The stability and chemical interactions are defined through the triplet nucleotides, and thus provide a differentiated reaction environment for repeated assembly of amino acids. The amino acids are matched to the interaction vessel based upon the structural and chemical correlation with the associated interaction vessel, of which there are 64 unique configurations.

Figure 5.13: After the amino acids are coupled through rotation out of the complex aided by the steroid molecule, a protein results whose characteristics are defined by the vessel from which it demonstrated stability. Thus, the process is completed and repeats as long as the strands remain open.

Chapter 6

Translating Decoded DNA

It was shown how amino acids, can be assembled in the reaction vessel according to its chemical and structural characteristics. In this chapter, I examine the sensitivity requirements of the reaction vessel in order to uniquely assign the amino acid to the reaction vessel. Here, I will discuss the feasibility of using the reaction vessel to differentiate amino acids, and mainly describe it from an inclusionary perspective, that is how the individual reaction vessels can support the intended amino acid. Some discussion of comparative nature is given to better understand the reaction vessel and a table for prediction of the molecules that would also reside in the vessel is examined.

6.1 Structural sensitivity of the interaction vessels

To evaluate the sensitivity of the interaction vessels to differentiate between amino acids, it is useful to list the potential differences of the structural and chemical characteristics. For a set of twenty amino acids, in order to get to a unique set, it is necessary to find at least twenty unique differences in the structural and chemical composition, which can be a combination of attributes. In studying the structural attributes, Table 6.1 indicates the structural differences among the different reaction vessels.

The most striking comparison is the size of the residue group of the amino acid. It can be seen in looking at a chart of the twenty amino acids that there is good correspondence between the assigned triplet and the associated reaction vessels. (Note, when performing this analysis, it is important to remember to swap out the complementary base pairs for a nominal UCGA table, since the approach here is to directly translate off of the nucleotide base, rather than go through the intermediary

		Amino Acid																			
	Metric	A	C	D	E	F	G	H	I	L	K	M	N	P	Q	R	S	T	V	W	Y
Structural	Residue Oriented to Project Down the Vessel	X	X	X	X	X	X	X	X	X	X	X	X	X	X	X	X	X	X	X	X
	Shorter Residue	X	X				X		X	X		X		X			X	X	X		
	Longer Residue			X	X	X		X			X		X		X	X				X	X
	Linear Residue	X	X								X	X					X				
	Symmetric with Ring Structure					X								X						X	X
	Compatibile with Weak Walls Unable to Retain Residue						X														
	Larger Element than C, O		X									X									
Chemical	Oxygen and Amino End Groups	X	X	X	X	X	X	X	X	X	X	X	X	X	X	X	X	X	X	X	X
	No Functional Group	X					X		X	X									X		
	Functional Group Interaction		X	X	X	X		X			X		X		X	X	X	X		X	X
	Oxygen Influence Higher Position								X			X					X		X		
	Oxygen Influence Medium Position								X	X		X	X		X			X			
	Third Base Influence			X	X			X					X		X	X					
	Base Influence			X	X			X		X			X		X	X					X

Figure 6.1: Detailed and refined structural and chemical differences of the reaction vessels formed by the nucleotide and steroid pairings define a number of possible configurations that define which amino acid residue is situated within the reaction vessel.

of mRNA). The shorter residue groups such as those of valine, alanine, threonine are associated with the bases that comprise the purine molecules, which will be closer to the edge since they are two rings, while the pyrimidine molecules contain only one ring.

Further structural issues include the strength of the sidewalls. At an extreme, the CCC pairing will be the weakest since there is little support, there are no functional groups on the sidewalls to assist with stabilization, and there is a long distance to the nucleotide base. It is thus the case that no amino acid with a residue can enter the reaction vessel because it is too unstable and will not allow sufficient time to interact with the candidate amino acid. The only amino acid that can interact with TTT would be glycine, with has a hydrogen element as a residue group.

The chemical interaction of the vessel are also important and in Table 6.1, the chemical interactions are evaluated. The chemical differences include the sidewall functional group, and its positioning. Other chemical features include the nucleotide base itself, and whether there are two or three functional groups, and the nature of those functional groups, whether it be a ketone or amino group. The chemical charge of the bases on the first and third nucleotides is less important, but can play a role in certain circumstances.

It is important to comprehend the specificity enabled by the first and second walls corresponding to the first and second nucleotides, and thus, the code should emphasize the first two letters because of these requirements. The third amino acid is necessary to support the second wall, but is not as important as the first two nucleotides, as it is only providing symmetry and support. Therefore, the third nucleotide would be considered to be necessary but not sufficient to define performance as it is on the exterior of the reaction vessel. This can be seen in the amino acid to nucleotide code, where the third nucleotide for some pairings is not influential at all, and for other interactions it plays a role. However, the third nucleotide is critical because it provides the wall from which the steroid molecule transporting the amino acid to the interaction vessel performs alignment and orientation. Thus, a strong result indicating the correctness of the approach is that the organization of the amino-acid to triplet nucleotide is explained in that the third nucleotide has less impact on the mapping of the amino acid to nucleotide, since it provides support and alignment, but not the actual wall for the association of the amino acid to the base.

Some unusual cases can be examined, such as the starting site, which has only one possibility. This can rationalized by considered the size of the SH group and its position within the candidate vessel, and the strength or flexibility of the side walls. This can be contrasted with the stop sites.

6.2 Translating codes

The inverted code is presented for short nucleotides in the second position in Table 6.1. Thus this works by given the interaction vessel and its neighbors, a prediction on the amino acid could be given in the inclusionary sense and as justification to support the developments. It can be seen that each of the amino acids can be rationalized to be assigned to a location on the table. More work is needed in this area, more modeling to determine the interaction of the amino acids with the reaction vessels. But to a first order approximation, there is agreement, particularly on size.

As size is a good first indicator of the interaction of the amino acid with the interaction vessel, Table 6.2 indicates the inversion of the characteristic with the longer residues of the amino acids, which are matched to the pyrimidine bases. There is general agreement based on size, of which glycine was discussed on how the instability of the reaction vessels precludes anything but glycine interacting with it.

In Figures 6.2, 6.3, 6.4, and 6.5, the decoding tables are shown that correlate the structure and chemistry of the interaction vessel with the corresponding amino acid. The justification for these amino acids is taken from Table 6.1 and 6.2, and the structural and chemical checklist. It would be possible to correlate a second table of amino acids that may compete effectively for the reaction sites. This can then be inverted to determine the best amino acid given a set of conditions corresponding to the interaction vessels.

AA	Reaction Vessel Requirements	Sequences
G	Weak walls prevent entry other than hydrogen,non-functional group	GCA,GCG,GCT,GCC
F	Symmetric, Functional groups, Strong walls to support aromatic ring	AAA,AAG
T	Functional Group positioned left low to attract OH	TGA,TGG,TGT,TGC
S	Functional Group positioned left higher to attract OH	AGA, AGG, AGT, AGC
M	Two functional groups to lodge large molecule with flexible walls	TAC
I	Charged walls at low level left, higher right	TAA,TAG,TAT
L	Strong walls, left charge positioned low	AAT,AAC,GAA,GAG,GAT,GAC
V	Weak left wall, no left functional group, two functional groups on base	CAA,CAG,CAT,CAC
A	Weak left wall, no left functional group, three functional groups on base	CGA, CGG,CGT,CGC
P	No charge, strong walls to support ring group, no functional groups	GCA,GCG,GCT,GCC

Table 6.1: For the purine bases, which are longer molecules because of its two rings, the length of the reaction vessel will limit the kind of molecules that can stably interact within the vessel while retaining connection with the exterior of the vessel. Thus, the differentiation of the amino acids for the shorter amino acids are compared.

6.3 Summary of Translation Table

The translation takes place through the structural and chemical sensitivity of the interaction vessel with the amino acid. Because of the projection of the vessel a first level of chemicals that resolve this are associated with the orientation of the amino acid, which is limited to L configuration as the D configuration will project in the opposite direction. The size of the residue is the next factor as it immediately separates the amino acids in to those that interact with purine and those that interact with pyrimidine. More detailed chemical and structural characteristics are then used to assign the twenty amino acids to the vessels. Using the inversion tables, it is possible to predict which interaction vessel, and thus triplet alphabet, corresponds to a candidate amino acid.

In Figure 6.6, an initial evaluation of the structure and chemical characteristics of the interaction vessel with the amino acid is indicated. The table indicates that there is a correlation in the physical size of the residue of the amino acid with regard

AA	Reaction Vessel Requirements	Sequences
R	Strong Walls, Uncharged walls, Three group interaction with C	CCA,CCG,CCT,CCC
R	Functional Wall on left for interaction	TCA,TCG
S	Functional Wall with O well positioned to attract OH	TCT,TCC
W	Some symmetry, stronger wall, some flexibility	ACT
C	Stable association environment with SH group, stronger walls,able to retract due to shortness	ACA,ACG
Y	Charged walls on both sides, and 2 functional groups	ATA,ATG
H	Strong walls, charge on right side wall	GTA,GTG
Q	Lots of charge and stronger walls	GTT,GTC
N	Lots of Charge but weak walls, though not as weak as K	CGA,CGG
K	Lots of Charge but weak walls hence linear group	TTT,TTA
ZW	Larger molecule with stability	STOP
ZC	Stronger larger molecule with stability	STOP

Table 6.2: For the pyrimidine bases, which are shorter than purines as there is only one ring, hence the amino acid residues need to be relatively longer in order to interact and secure successfully while retaining connection with the exterior of the vessel. Thus, the comparison of the longer amino acids are compared relatively to determine their differentiation with respect to the pyrimidine bases.

to the first two columns being shorter than the third and fourth column. This is correlated to the size of the base where the first two columns correspond to a purine molecule which is two rings in comparison to the pyrimidine molecule which has only one ring. Thus longer molecules will be needed to interact and stabilize with the base than those of the first two columns. Because of the longer distance, there is increased sensitivity to the strength of the walls, in particular the size of the base of the third nucleotide. A shorter base of the third nucleotide position will allow the wall to flex more and can lead to larger residues, which is the case in comparing Group A and B residues. Furthermore, glycine is the shortest amino acid, and the reason for that is that the interaction vessel is unstable, and nothing other than hydrogen residue can associate with it. This instability is seen as there are no functional groups on either wall, it is a long distance from the entry point, and the size of the base of the first nucleotide is the smaller purine. Other modes of inclusion can be noted for this set of amino acids relative to the physical dimension and chemical characteristics of the

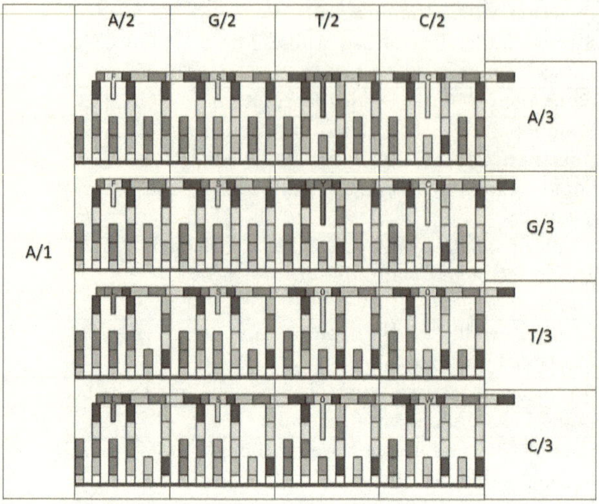

Figure 6.2: Mapping the construct to an amino acid for adenine as the leading nucleotide, the translation of the triplet structure to assemble an amino acid based on structural and chemical interactions over the second nucleotide. Note that the table can be organized on size relative to the second nucleotide, that is the columns of the table. This is consistent with the length of the interaction vessel being the first differentiation factor. In these set of four tables, it is important to see that an amino acid is assigned to each of the sixty-four interaction vessels, which can be seen as the one letter stamped into the colorized amino acid. These are selected based on the tables of chemical and structure correlations, and on reflection of the amino acid to triplet nucleotide mapping.

interaction vessel.

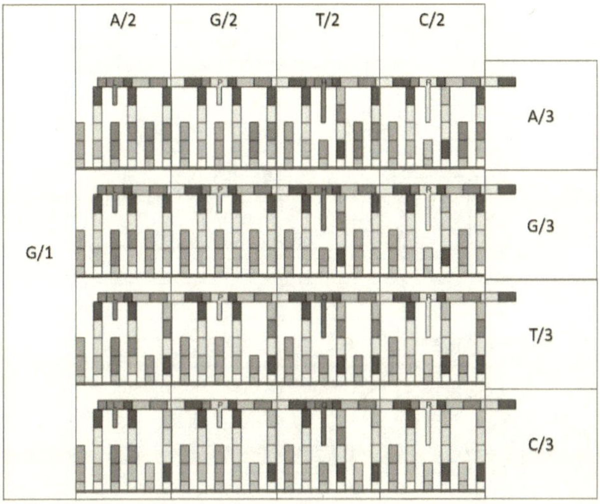

Figure 6.3: With guanine as the leading candidate, the decodingof the triplet to the association of an amino acid is indicated. The amino acids have different colors to indicate different amino acids. As a reminder, each rectangle relates to a different ring structure, whether on the nucleotide or the steroid molecule.

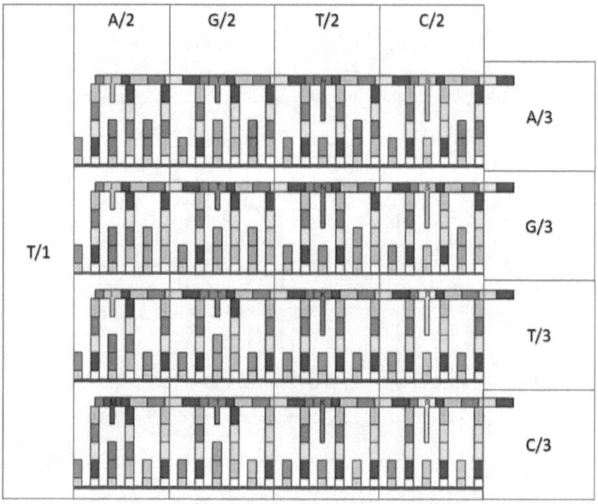

Figure 6.4: The decoding of the triplet with thymine as the first nucleotide, and as noted with the others, the second nucleotide is a major differentiating factor since it defines the length of the interaction vessel. In this section, the start sequence methionine is located at an individual site, which can be rationalized through examination of the chemical structures.

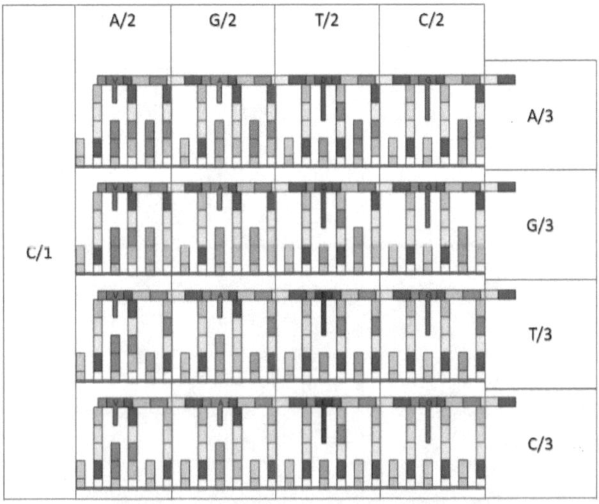

Figure 6.5: Completing a total of sixty-four interaction sites for twenty amino acids, cytosine as the leading nucleotide is shown for the translation table. Of interest is glycine, which is the shortest amino acid residue, but it is present in the longest reaction chamber, but the chamber is very weak since there are no functional groups on its sidewalls, the support on the side walls is weak, and the distance is long. Thus longer molecules can not enter successfully, and glycine does not have these problems since its residue group is a hydrogen element.

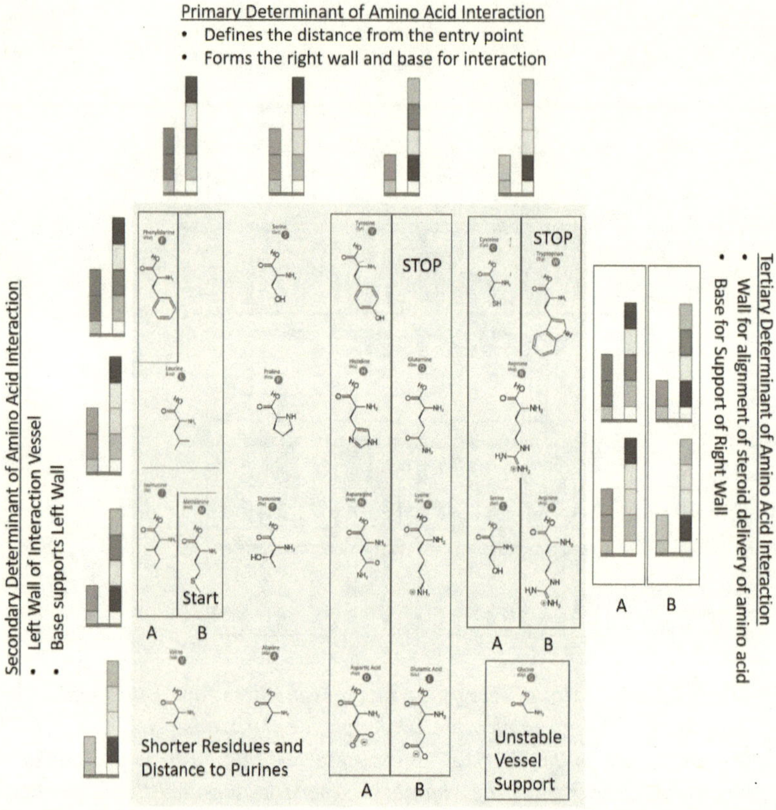

Figure 6.6: An evaluation of the translation of the interaction vessel and its differentiation with amino acid selection and assembly. This chart is the first to ever indicate the mapping of the structural relationship of the DNA nucleotides to the amino acids through an interaction vessel constructed as two walls of four-ringed molecules, steroid molecules, with a third wall for aligning and docking another four-ringed molecule that associates the amino acid to the base of the second nucleotide.

Chapter 7

Example of DNA Encoded, Transmitted, Decoded, and Translated

Applying the results for the simulation and design of DNA for producing molecular structures, and in simulating what would be the origin of DNA and its transcriptional mechanisms, this section presents the chemical structures associated with encoding DNA, transmitting the information, and then decoding the information so as to obtain a desired result, which in this case is a protein structure. In addition, the inverse direction is discussed, in which the desired structure is specified, and then the encoded structure is synthesized, which would result in the desired structure when implemented.

Rather than specify the starting triplet of nucleotides to serve as an initiation site to enable strand separation and then an upstream transcription point, a starting triplet is directly selected to illustrate the encoding, transmission, decoding and translation, I will randomly select molecular orientations for 25 pairs; hence a total of 50 steroid molecules will serve as the starting set, which will result in 25 nucleotides and 25 corresponding steroid molecules. The decoding of the system into a series of interaction vessels whose structure and chemical nature will define the associated amino acid, which is selected from the chart of Figure 6.6.

7.1 Encoded DNA

For this example, the following twenty-five nucleotide-steroid pairs, resulting in a total of 50, were selected by an independently, which is from a random number generator. The actual sequence of the 50 pairs is listed in the Appendix, and is illustrated in the graphical form of Figure 7.1. The specification of this encoded sets forth the entire synthesis process, and the formation of the chain of interaction vessel comprised of nucleotides interleaved with steroid molecules, which will produce a resultant amino acid chain and protein.

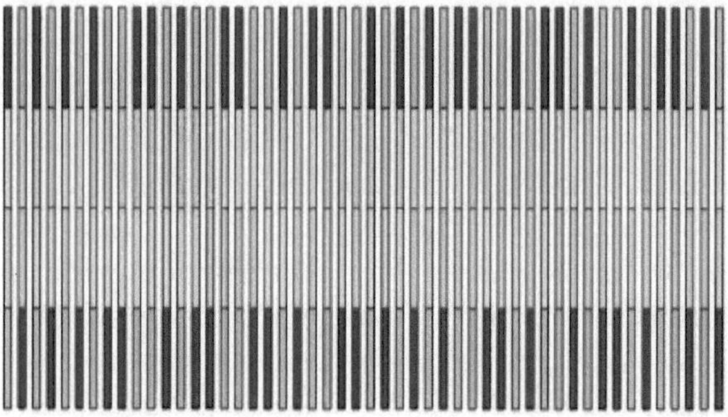

Figure 7.1: The sequence of the orientation of the pairs of pairs of steroid molecules is indicated for the twenty five pairs. This is a representation of a physical structure, that is an arrangement of tetra-ring molecules. The complex is comprised of pairs of molecules connected by hydrogen bonding on either end. Depending upon the orientation of the four rings comprising the steroid molecule, of which each ring is represented in the figure by a rectangle, the resultant nucleotide sequence is encoded within this structure.

As there are 25 pairs of coupled molecules, there will be 24 inter-pair intermolecular couplings. If the last molecule is considered to be coupled with the first molecule, there will be twenty-five inter-pair couplings, and that is the route that will be taken. Recall that each pair is coupled at the ends through hydrogen bonding, and then each pair relative to the next pair can have its oxygen molecule on the hexa-ring at the end either adjacent or opposite to that of the next pair. This position will define the nature of the nucleotide, whether it will become A-T/T-A or G-C/C-G. then the hydrogen bond coupling for the twenty-fifth pair could be determined. Applying the encoding Table 3.2, the hydrogen bonding of inter-pairs is indicated as:

$$H_{bond} = [01111100101010100010001100] \tag{7.1}$$

Thus, this step of pairing pairs of tetra-ring molecules with functional oxygen groups has chemically encoded a structure that will ultimately become a DNA nucleotide sequence as well as its paired steroid molecules that will enable its further function to put its information content into motion. The message as a chemical structure is thereby fully encoded and ready to be transmitted. It is noted that at this stage the encoding of the molecule can be deciphered mathematically to determine the nucleotide sequence as long as the hydrogen bonding sequence is determined. From the orientation of the rings and the resultant hydrogen bond arrangement, the nucleotide sequence for this example is:

$$\mathtt{DNA_{example}} = \mathtt{GAAATAGCACTGTCTCGTCGGTTGG} \tag{7.2}$$

7.2 Transmitted DNA

To prepare the DNA and steroid co-enzymatic complex for transmission, it is arranged with an attachment of a phosphodiester backbone in which the DNA nucleotides by definition are covalently attached. The steroid molecules are free, but are stabilized by hydrogen bonding to the phosphor group. This is represented in Figure 7.2.

This co-enzymatic complex defines a set of DNA nucleotides that are synthesized with the steroid molecules. The DNA nucleotides form a connected base 4 construct, and the steroid molecules take on two forms, either with or without a mid-molecule hydroxyl group. The steps of the synthesis are represented in the appendix, and using the procedures described, the formation of the unified complex prior to the double helix is given in Figure 7.3. This structure includes the nucleotide pairing and its associated steroid molecule. Some of the DNA nucleotides are not oriented in a position compatible with the double helix as we understand it: that is, on average half of the purines will face forward, rather than backward; and half of the pyrimidines will face backwards, instead of forward, when viewed from a common strand.

To form the double helix, it is necessary to rotate the DNA nucleotides into position, which will thereby release the steroid molecules from the complex. After nucleotide rotation, each purine is aligned in the same direction, and each pyrimidine is aligned in the same direction. Interestingly, the purine are aligned as BFBF and the pyrimidine are aligned FBFB, which will balance the configuration. On average, this

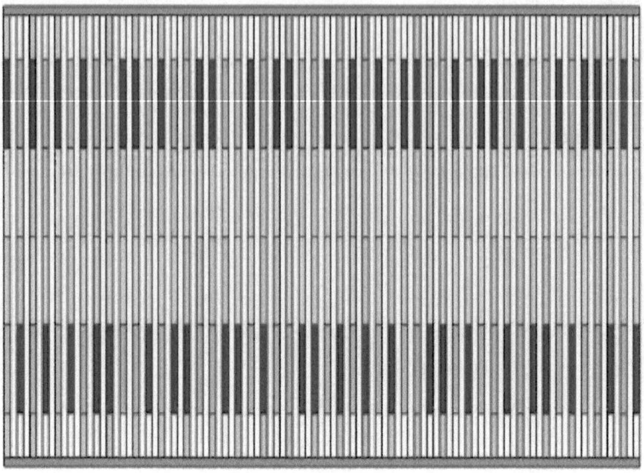

Figure 7.2: For the example, after association with the phosphodiester backbone the nucleotides are attached, but still comprise a steroid molecule, and the accompanying steroid molecule are stabilized at the ends by hydrogen bonding, and internally by aromatic stabilization.

will require "flipping" half of the synthesized nucleotides to obtain the stable double helix. This rotation frees the steroid molecules, which can then be used immediately to open the strands for further processing, including amino acid assembly. The nucleotides which would have to flip can be determined mathematically, and that sequence is given by for this example as

$$\mathtt{O_{flip}} = [0101011100110011100001000] \tag{7.3}$$

. in which $\mathtt{O_{flip}}$ indicates whether the nucleotide flipped in order to arrive at the final configuration. Thus there were 11 nucleotides that rotated into position.

In Figure 7.4 the unit after the formation of the double helix is indicated in which formation of the double helix is achieved by rotating the out of phase nucleotides, and ejecting the steroid molecules. The strands are thus available for association with steroid molecules such that opening can initiate transcription, if the strands require opening. Thus there is an instantaneous life function with the release of the steroid molecules, which interact directly with DNA to induce genetic function and distribution of the information content.

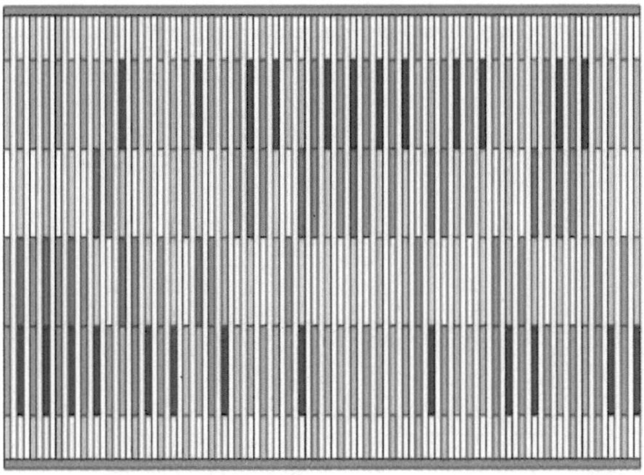

Figure 7.3: The nucleotides are synthesized, connected by hydrogen bonding, and the steroid molecules are in place with two types, either with oxygen groups for intermolecular hydrogen bonding, or without; both have planar arrangements. For the example, the synthesized co-enzymatic complex right before it disassociates into a double helix.

7.3 Decoding DNA

After receiving the message as a separated strand of DNA, the next step is to perform error checking by a bitwise comparison with its associated steroid molecule in populating the strand at the location of the phosphor molecule on the phosphodiester backbone. The result of the nucleotide matched with a steroid molecule in its adjacent location is indicated in Figure 7.5. In addition to error checking, this configuration prepares the overall structure for amino acid assembly by delivery of the amino acid by the workhorse steroid molecule, which also forms the ceiling of the overall structure.

In Figure 7.6, the final configuration in which the matching of the candidate amino acids within the assembly vessel. The beam on top is comprised of steroid molecules which transported the candidate amino acids to the interaction vessel. If unstable, for example, the residue of the amino acid is too big or too small to fit within the interaction vessel or fails to bind because of a lack of functional modes, it

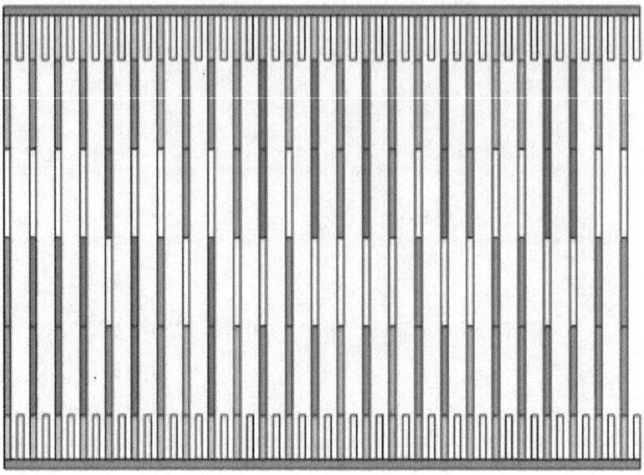

Figure 7.4: Formation of the double helix by rotating the out of phase nucleotides, and ejecting the steroid molecules. The strands are thus available for association with steroid molecules such that opening can initiate transcription.

moves onto the next candidate. If the fit between the residue and the chemical and structural characteristic of the interaction vessel works, then a stabilization process occurs in which there is ample time to connect with the adjacent steroid complex, and so forth until the chain of amino acids gets to the point where peptide bonds are a lower energy configuration, rather than balanced in a long chain of steroid molecules.

The chart is used to match the reaction vessel to the nucleotide. The nucleotide sequence is thus AGCTCCT.

7.4 Translated DNA

To determine the amino acid sequence, the chemical and structural information of the vessel is affiliated with an amino acid that is compatible. If the second base of the triplet comprising the interaction vessel is a pyrimidine molecule, that is cytosine or thymine, the distance to the entry point is longer, and hence an amino acid with a longer residue is necessary. For the purine molecules at the base of the

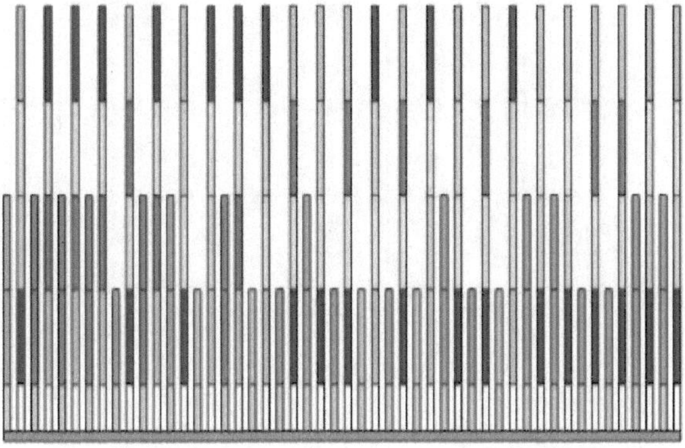

Figure 7.5: Separation of the strand, ready for interaction with the external environment to generate proteins by assembly of amino acids. Note that the steroid molecules when positioned in alignment with the nucleotide provides a uniform base of interaction at the area which was separated.

interaction vessel, the opposite is true, and shorter residues would be appropriate; otherwise coupling with the steroid hormone is not possible at the entry point. For a starting point of the first triple as AAA, the comparative charts indicate that structural symmetry is achieved at both walls through association with Phenylalanine (F) will be the first amino acid. The next nucleotide set is TAG at the has a purine molecule as the second molecule, and thus a shorter residue is needed. There are also two functional groups on the wall as indicated in the structural chart. This would favor interaction with an amino acid that has a functional group; however the adenine base is apparently bound to the wall functional group, and thus a neutral molecule is enabled with methyl projections that must be consistent with the internal interactions of the cell, including those of the wall. The choice is thus isoleucine (I). The general strategy of translating the interaction vessel construction to amino acid is indicated in the translation chart of Figure 6.6. Using this chart to continue the analysis for the interaction vessels associated with the nucleotide sequence of AAATAGCACTGTCTCGTCGGT. After 7 amino acids are assembled in the protein and connected, which as indicated in Figure 7.7, it results in the protein structure

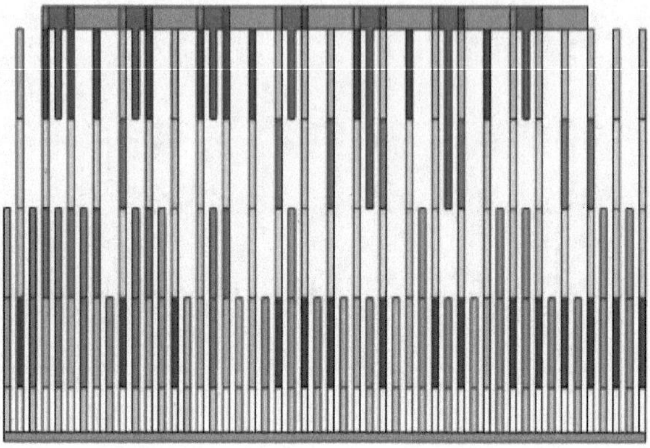

Figure 7.6: The final configuration in which the matching of the candidate amino acids within the assembly vessel. The ceiling beam on top is comprised of steroid molecules which transport the candidate amino acids to the interaction vessel. If unstable, for example, too big or too small, it moves onto the next candidate. If works, then the stabilization process connects to the adjacent steroid complex, and so forth until the chain is complete. (In this representation, the color coding of the steroid molecule on the roof of the interaction vessel is not indicated.)

FIVTEQP.

7.5 Inverse Feedback

In addition to the forward direction of providing an encoding sequence, which might be the case in a creation mode; the other direction, the inverse or synthesis direction, is also easily determined. In this situation, the desired nucleotide sequence or the amino acid sequence is given, and the problem is to encode it such that if the same methods were applied, it would produce the specified nucleotide sequence or the amino acid sequence. For example, in this study the first nucleotide is G, which would be either PFBK or PBFK depending on the sequence of a preceding steroid molecule, which for this problem was considered as the fifty molecule, KFBP; thus

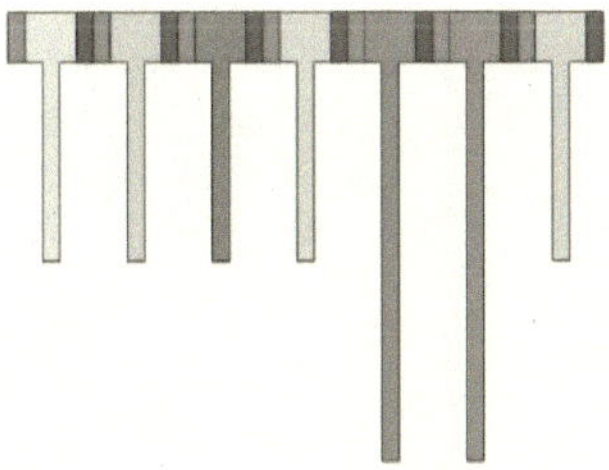

Figure 7.7: The final representation of the protein as FIVTEQP after identification of the stabilizing amino acid and its removal from the interaction vessel. The shorter molecules which align to the pyrimidine bases C and T are alongside the longer molecules which align to the purine bases G and C.

the first molecule would be PFBK, since it would not have hydrogen bonding, and thus there is only one choice from Table 3.2. This corresponds to the steroid orientation KFBP. Thus, to achieve the next nucleotide, A, would involve selecting the appropriate sequence using Table 3.2. It requires hydrogen bonding since A-T only has two internal hydrogen bonds and three are required, and A is a pyrimidine, thus it must start with P. Thus, the choice is PBFK. This process is repeated until the encoding sequence is finished for the DNA-Steroid complex. The exact result would be produced of course since the encoding sequence is the same.

For this protein, the interaction vessels indicated that the protein sequence was FIVTEQP. If only this information was provided, then there would be several choices on the nucleotide sequence. In this case, one would be selected, which would result in a nucleotide sequence. Since there is a one-to-one mapping between nucleotide and the encoding sequence, a set of reaction vessels would result that would match exactly the original protein sequence. Hence, in either case, nucleotide or protein specification, the synthesis direction is feasible. Note that this also enables a feedback loop, which implies that the resultant synthesized protein can be fed back to influence the encoded nucleotide sequence, which after processing results in a different protein, perhaps with better, or at least new, properties.

Chapter 8

Replication

Replication is necessary for life function. The double helix form of DNA, which was formed by the simultaneous release from a unified complex with steroid molecules, can serve as a template from which subsequent copies can be made. Despite its use as a replication template, it is important to note that the double helix has less information content than the unified complex due to the rotation of the base pairs into a lower energy state. Thus, this chapter will provide information on these replication topics:

- Method to produce replicated strands from the double helix using steroid molecules as the intermediary structure.

- Re-creating the true master image, which was the unified complex that produced both structure and function.

Thus, it is important to indicate that the images produced from the double helix, which is a less informative master than the unified complex, will only contain structural information. The true master image, which contained both structural and functional information, will thus need to be recreated from the double helix, if both structure and function is utilized. All copies off of the double helix, which is derived from the unified complex, are structural-only reproductions. The double-helix images do not impart functionality, because there is a loss of information due to the rotation of the purine and pyrimidines into low energy state positions, thereby losing there representative orientation. However, it is possible to recreate that orientation, and that is one area that will be investigated herein.

8.1 Replicating the sub-master double helix DNA

The replication of the double helix is a fairly straightforward process with the use of the steroid molecules as a template to guide the complementary base pair into position on the single strand DNA. After the assembly of the molecules along the single strand of DNA, and the template positioning, the growth of the phosphodiester sugar backbone will result in the replication of the double helix. Thus with strand separation and replication off of each strand, a doubling of the original material results. The steroid molecule is released as the template during the replication process, and is thereby available for continued replication.

In Figures 8.1(a) - (d) for A, G, C and T, the molecular model is indicated for the association of the complementary base pair with a steroid molecule that is configured for that particular base pair. During the transition of the unified complex into the double helix and the steroid molecule, two types were produced. One associated with the C-G pairing, which did not contain a mid-molecule hydroxyl group, and for the A-T pairing, which does have a mid-molecule hydroxyl group. Figures 8.1(e) - (h), the van der Waals spheres indicate that the planarity, spacing is appropriate for mirroring the steroid and deploying it as a template. Figures 8.1(i)-(l) the representation of the steroid and complementary base pair is indicated in the key format that will be used for continued analysis, showing the ring structure and functional groups comprising the steroid and base pairs, as well as its orientation. The first step for replication involves the separation into the single strand DNA. Then to prepare the strand for replication, it is filled with complementary base pairs and steroid template holding it together.

To develop the procedure for replication, an example is presented for the single strand DNA sequence of `TAGTATGCGGTTCCGACGTCTA` as indicated in Figure 8.2(a). The separation of the double helix into a single strand can take place through the binding of the steroid molecule onto the double helix, which will locally separate the strands forming a small bubble about the binding site. Additional binding will form numerous bubbles along the strands, resulting ultimately in the complete separation of the strands, some of which is driven by the incorporation of the steroid molecules within the section for direct transcription and translation, which can also be utilized . After filling the strands with the steroid molecule will prepare for replication, as indicated in Figure 8.2(b), after association with the complementary base pair.

In Figures 8.3(a) - (d), the example is further studied to indicate the progression of the replication process after the strands are separated and the sites are filled with the template and base pair. The filling process can occur while the replication process is proceeding. The initial step is the start of the fabrication of the phosphodiester

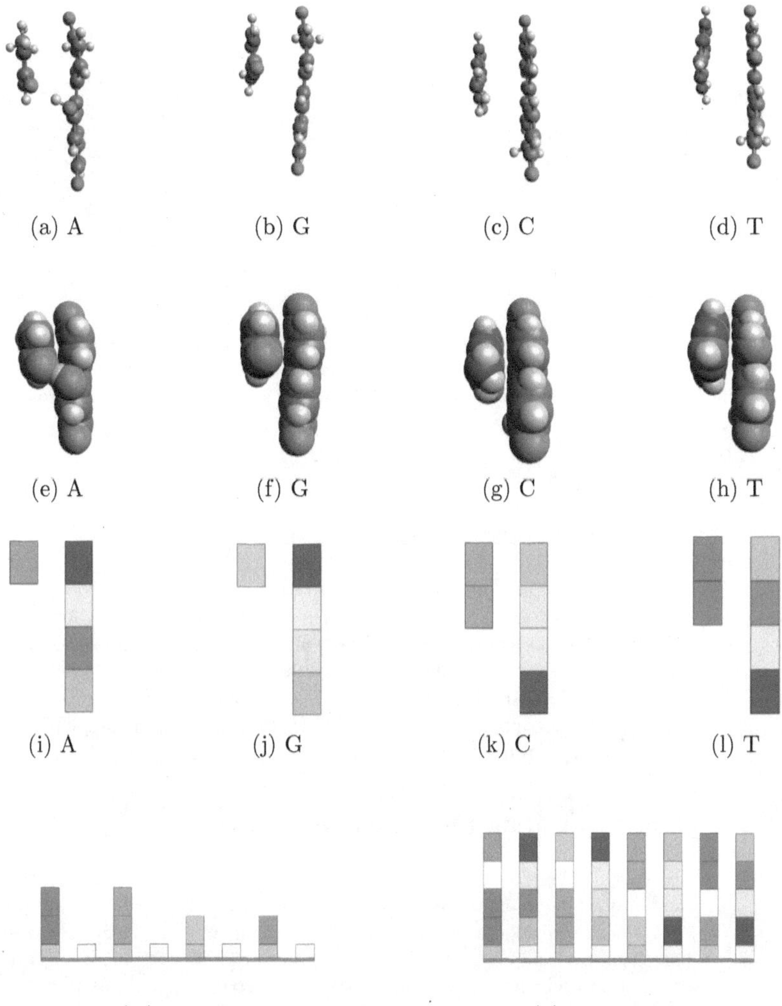

(a) A (b) G (c) C (d) T

(e) A (f) G (g) C (h) T

(i) A (j) G (k) C (l) T

(m) AGCT (n) AGCT Filled

Figure 8.1: (a)-(l) Representations of the planarity and alignment of steroid and complementary base pair; (m) single strand DNA; (n) ready for replication.

backbone, which enables the release of the steroid template as the double helix is formed, as indicated in Figure 8.3(b). This process continues until the overall process is completed. Thus, an efficient mechanism of replication is enabled through

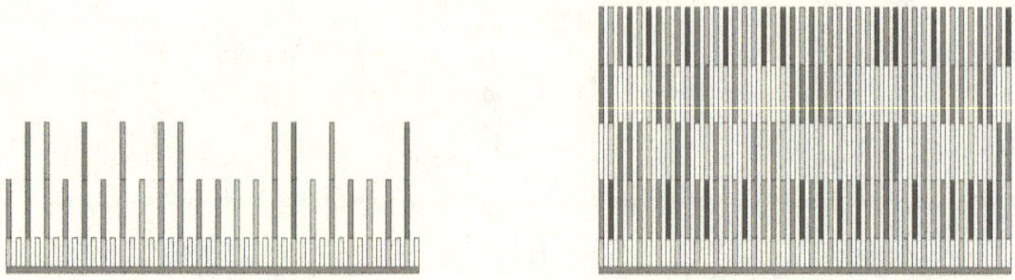

(a) (b)

Figure 8.2: (a) The example strand of DNA used in this application in replication. The sequence `TAGTATGCGGTTCCGACGTCTA` *is used. (b) After filling with the complementary base pairs and stabilizing the overall structure with steroid molecules.*

deploying the steroid molecule as a template.

8.2 Error Correction and Repair

In addition to replicating off of a single strand, the replication procedure of Figure 8.3 can be used to correct for errors in the double-stranded DNA. The steroid molecule together with the complementary base pair are aligned within the damaged zone, whether comprised of a single or many nucleotides using the technique presented in Figure 8.1 at the damaged zone. The phosphodiester sugar backbone is then built along the edge in the manner shown in Figure 8.3 to complete the correction procedure. This procedure is graphically indicated in Figure 8.4.

8.3 Replicating the True Master Image

To replicate the Master image, which contains much more information than the DNA double helix, is an enormous task. The issue is that the purine and pyrimidine bases comprising the double helix are all facing the same respective direction, purines back, and pyrimidines forward. However, the true Master image contained different orientations, which was lost when distributing structure and function, in the form of a released steroid molecule and the DNA double helix. Thus, in order to replicate the true Master image, it is necessary to recreate the original orientation from the double

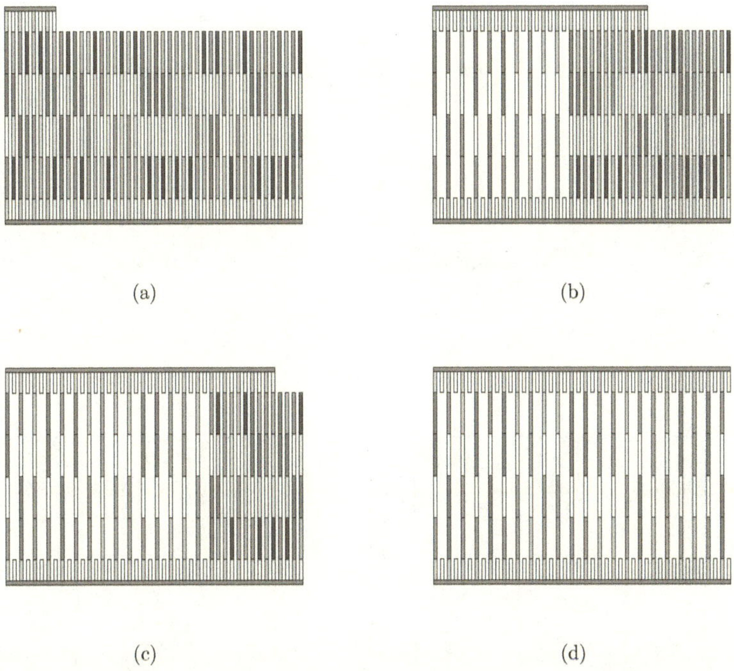

Figure 8.3: The procedure to produce a replicated double helix DNA from a single-strand of it. Thus a doubling in capacity from the original DNA. (a) the phosphodiester backbone begins synthesis attaching to the nucleotides; (b) after some nucleotides the double helix will be twisted into, and the steroids are ejected; (c) this process continues to form the double helix; (d) all the way to the end, the phosphodiester backbone is completed.

helix. And even in this event, although it is possible to determine the orientation, since the very first steroid orientation is unknown, the phase may be completely opposite from what was in the original. However, if one was to recreate the true Master image, it is possible to induce not only structure, but also function, and therefore create a critical life function.

To create the true master image from the double helix, it needs to be determined

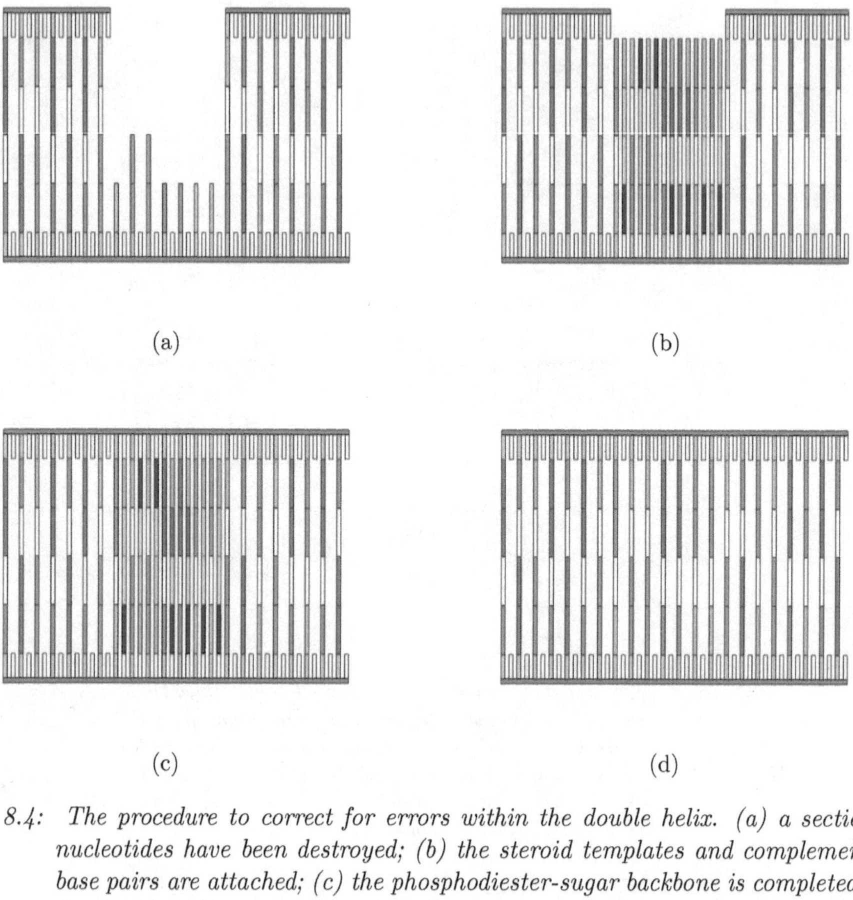

(a) (b) (c) (d)

Figure 8.4: The procedure to correct for errors within the double helix. (a) a section of nucleotides have been destroyed; (b) the steroid templates and complementary base pairs are attached; (c) the phosphodiester-sugar backbone is completed; (d) the double helix re-forms to eject the steroid molecules.

the original orientation of each base pair. Fortunately, this can be determined, and needs to be started from the first position and move in one direction. While the identification of the first nucleotide is known, its original orientation is not known, since it is defined by the proceeding steroid molecule, which is lost during the creation of a life function by releasing the double helix and the steroid molecules. Thus there are two solutions to the true master image, and they are completely out of phase, for each and every nucleotide. Thus, two master images would be needed, one of which will be the true master image, and the other will be its complete opposite. Both will yield the same double helix.

In Table 8.1, an example is given of the steps that are necessary to recreate the master image from the double helix. The original starting material is used for these developments. Using the original starting material, the first step involves mirroring the first nucleotide, then the original steroid molecular structure is brought in place in between the first and second nucleotide, which hydrogen bonds to the first molecule, while orienting the second nucleotide for correct hydrogen bonding. Then, the first nucleotide needs to move out of the way, and the first and second original starting material need to move upwards, to allow a third molecule of the starting material to mirror the newly oriented second nucleotide. This process then repeats for the entire chain. The result is a new unified complex of the True Master Image. The phase will be in accordance with the starting phase of nucleotide one.

Master	Nucleotide	Observable	Step 1 Mirror	Step 2 Insert	Step 3 Remove	Step 4 Mirror	Step 5 Insert	Step 6 Remove	Step 7 Mirror	Step 8 Insert	Step 9 Remove
PBFK			KBFP	KBFP	KBFP	KBFP	KBFP	KBFP	KBFP	KBFP	
KBFP	C	KB	KB	KB	PBFK	PBFK	PBFK	PBFK	PBFK	PBFK	KBFP
PBFK				PBFK		KFBP	KFBP	KFBP	KFBP	KFBP	PBFK
KBFP	T	KB	KB	KF	KF	KF	KF	PFBK	PFBK	PFBK	KBFP
PFBK						KBFP	PFBK		PBFK	PBFK	PFBK
PBFK	G	PF	PF	PF	PF	PF	PB	PB	PB	PB	PBFK
KBFP										KBFP	KBFP

Table 8.1: This is the procedure to produce the True Master Image of the Unified Complex based only on the observable nucleotide sequence of the single strand of DNA. It involves the mirroring of the nucleotide by the original material comprising the True Master Image, followed by the insertion of a steroid molecule of the original material. Through hydrogen bonding, the next nucleotide is properly oriented during the insertion process. The base pair on the single strand is then able to shift away from the complex, removed. These three steps are continued until the replication process is complete.

8.4 Reproduction hierarchy

The hierarchy of reproduction is consistent with the notion that from a unified complex, both structure and function were produced for life function. After life started, it was only necessary to reproduce structure, since function was available. This implies that there was sufficient quantities of molecular constructs from which function could go forward. In less abstract terms, structure was the double helix and function was the steroid molecules. The steroid molecules can be used to directly reproduce off of a single strand of the double helix without any loss of information content of the double helix. However, there is a loss of information of the unified complex, which contained encoded information on the steroid molecules as functional constructs. But since the steroid molecules are already generated, there further usage would require the ability to replicate the molecule in sufficient quantities, or at least use a similar molecule to accomplish its task in transcription/translation and replication.

Chapter 9

Genetic Rearrangement

To express genetic information after translation of the complex comprising the steroid molecules and DNA, several types of splicing strategies are available. This occurs to increase the complexity of the protein derived, as well as confer optimal properties in subsequent replication steps, which is observed in comparing exons and introns in eukaryotes, while prokaryotes tend to only have exons. The types of splicing strategies for the steroid-DNA complex utilizing replication techniques, include: (1) differential splicing of genetic constructs to generate a variety of proteins; (2) rearrangement of exons during replication by removal of introns and obtaining a larger, combined exon. Other areas of interaction, such as the defect generation misplacement of an exon inserted into an intron, or in the genetic rearrangement by recombination would also appear feasible under certain conditions.

9.1 Splicing Exons

To generate a variety of proteins off of the same strand, differential splicing is a mechanism in which exons connect during the translation process to create longer chains of proteins and arranged in different segments. While the mechanisms to produce alternative splicing schemes are known with modern-day mRNA, in this section, I show that it is also possible to perform differential splicing, connecting series of exons, through the use of the steroid, single strand of DNA, and thus this process can occur with the creation of life function, when the first DNA and steroid molecules are produced.

An example for the sequence `TGTACATGATCTCCGACGGGTACAAAATCGCG` is represented in Figure 9.1, which consists of the exon 1 `TACATGATC` and exon 2 `TACAAAATC` that contain a start and stop nucleotide triplet and an amino acid, `F` for exon 1 and

Y for exon 2. In this case, the splice is shown for exon 1 to exon2, which would occur by either the physical distance reduction by motion of the strand between the two sections that are covered by steroids, bypassing the intron section that does not contain a steroid ceiling, attaching F of exon 1 to M of exon 2, to result in the protein MFMY. This the differential splicing is achieved by the mechanism of deploying steroids to achieve functional genetic expression.

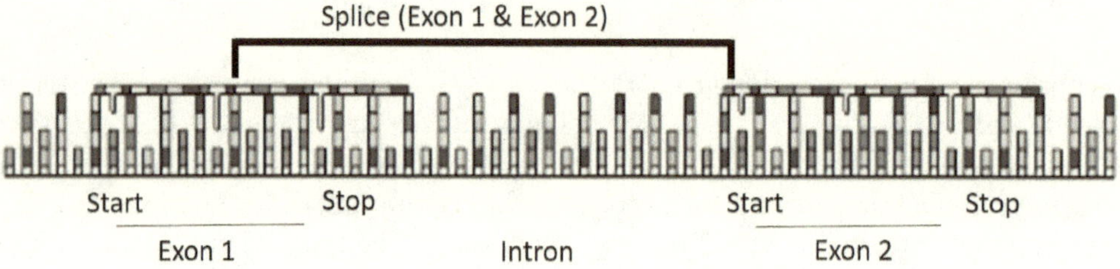

Figure 9.1: Splicing two exons together using the steroid molecule as the intermediary. In areas where the start and stop mechanics are not available, it is possible to splice the exons together. The Stop site is immobile due to the strong interaction of the amino acid (whose structure can only be assumed since it is non-expressive) with the interaction vessel. Thus, the last expressed amino acid is removed from its position and interacts with the Start amino acid of the next exon. This forms a longer protein structure, and increases the variety of proteins which can be generated.

9.2 Removing Introns from DNA Strands

During the replication process, it is possible to remove introns, and thus deploy only exons in the replicated strand. Thus, greater lengths of proteins can be built. This can happen if replication takes place soon concurrently with translation, in which a splicing event is taking place. If it occurs significantly afterwards, unless the strands are well stabilized connecting the two exons, the opportunity for bridging will be lost, at least on this particularly replication phase. This technique may be the reason behind prokaryotes generally comprised of a continuous strand of exons without introns. However, without introns there may be a loss in feedback control over individual protein segments. Figure 9.2(a) and (b) indicates this procedure. Two separate strands of DNA will result in the next replication cycle.

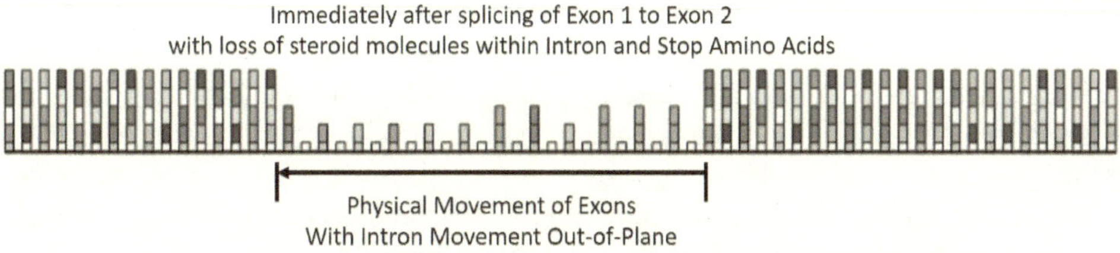

(a) Immediately after Transcription

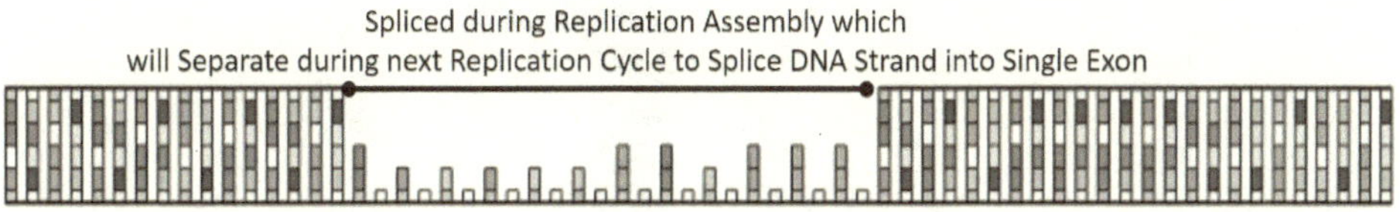

(b) Skipping Intron during Replication

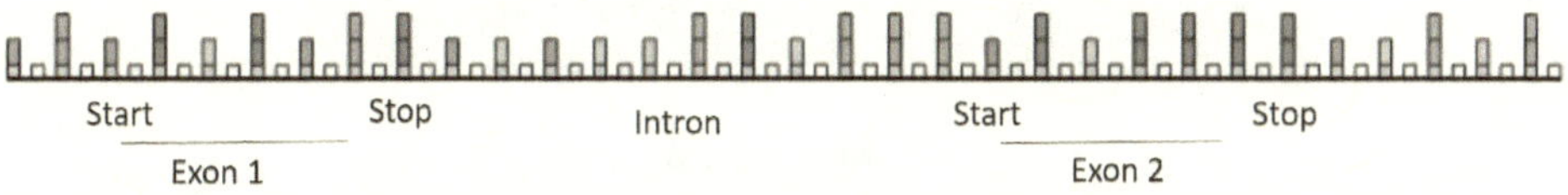

(c) Strand 1: After Next Replication Cycle

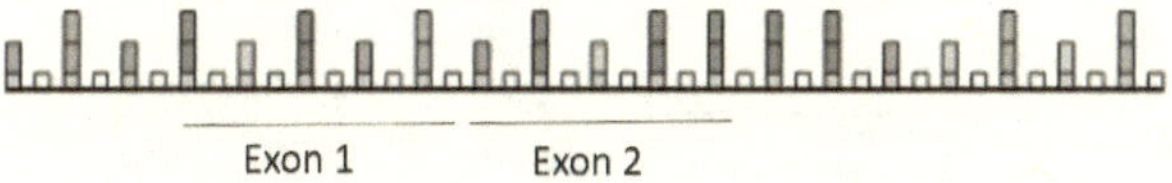

(d) Strand 2: After Next Replication Cycle

Figure 9.2: (a) immediately after splicing Exon 1 to Exon 2, and if the steroid molecules are detached from the Intron due to the motion of Exon 2 towards Exon 1, there is an opportunity to combine Exon 1 and Exon 2 into a single exon during the replication cycle. (b) If the timing is fast enough, the two exons can become spliced together during the replication cycle, of which the single strand DNA will separate during the next replication cycle. The original DNA containing both exons separated by an intron will remain in place (c), and the single strand of combined exon without the intron will result (d).

9.3 Gene Insertion and Recombinant DNA

There are other possibilities using this strategy for rearranging the genetic composition. This includes exon insertion in an existing exon, which can occur during the replication phase. In addition, during the replication phase, genetic recombination of DNA can occur. Both of these rearrangement processes occur by the strands coming in proximity to the replicating strand, and the transition from one part of the strand to the other during the phosphodiester backbone procedure guided by the steroid templates. A spacing model for recombinant DNA is shown by arranging steroid models in various crystals to show that the replication direction can be altered when implemented in conjunction with nucleotide base pairs.

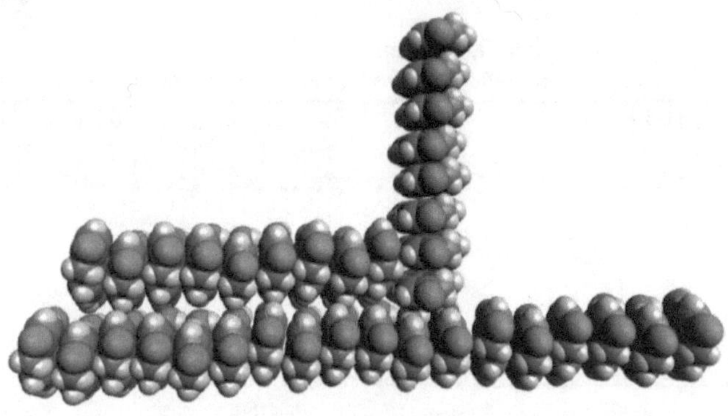

(a) Intersected

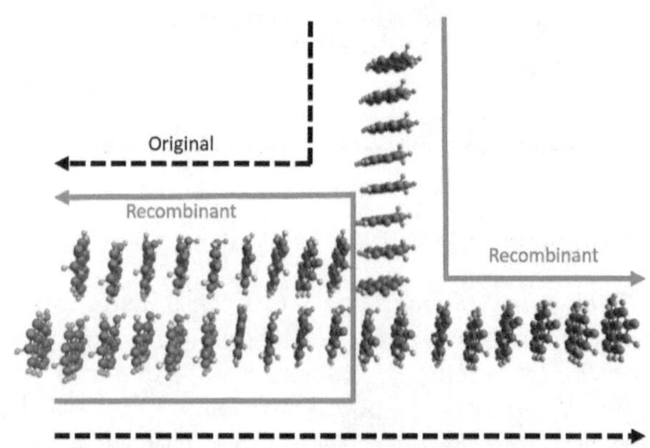

(b) Recombined

Figure 9.3: A spacing model to indicate feasibility of recombinant DNA forming through jumping onto an intersecting strand,

Chapter 10

Discussion

To impart a digital code within molecular structures, intermolecular coupling and intermolecular orientation were deployed to encode a structure that could also enable a co-enzymatic feature. This feature was used transmit the encoded digital representation that comprised two parts: one that contained the information (DNA) and the other part that enabled its function (steroid molecule) for decoding and translation. For decoding, which included bit error checking, a structure was formed by the steroid comprised of an interaction vessel. Moreover, it formed the ceiling of the interaction vessel, blocked off unnecessary zones, brought the amino acid to the structure, and participated in its formation as a protein structure. Both structure and function of a digital communications system were enabled through the association of pairs arranged within a unified complex.

The results are accurate and precise in characterizing the dimensions of the resultant interaction vessel comprised of steroid molecules serving as walls, as well as attached through the ceiling, with one portion open for the insertion of an amino acid. The centering of the interaction around the second nucleotide is consistent in analyzing the columns of the typical mapping of nucleotide triplet to amino acid.

As a valid solution describing the origin of DNA needs to be able to provide detailed answers to questions on its structure, the approach developed here provides such answers both with regard to DNA nucleotides as well as steroid hormones, and even the development of proteins comprised of amino acids. For example, questions that can be addressed are those that pertain to the most basic of which as to how the base four code is fundamentally defined, extending to the most detailed, such as a quantifiable metric associated with the arc distance between nucleotide pairs, which correlates with the pitch of the stacked basis molecules, both approximately seven angstroms. Moreover, there is agreement on the structural configuration, such as the rotation required to orient the nucleotides in correspondence to the present-day the double helix, which thereby enables a driving force to release the co-synthesized steroid molecules, connected with the prebiotic encoded amino acids. Remarkably, there are four orientations of one pair of steroid molecules to a second pair of orientations, thereby defining a base 4 code to DNA when formed as a sequential stack. There is a match in the structural biology of the steroid molecule and steroid hormones with paired DNA nucleotides: It has been shown that the match is perfect and undeniable, formed by a trace of precisely seventeen carbon, oxygen and nitrogen elements of nucleotide pairs. This is unique discovery brought forth for the first time by this research. And from this structural match, the mechanisms for the origin of DNA nucleotides and its related type of steroid molecules has been derived.

Early in the process, I have shown that the "code of life" is defined. This is a key step and insight, in which there is co-synthesis intermolecular development of the information content and the capability to read it and use it. Moreover, the co-development of the DNA as target and the steroid molecule, and their simultaneous release and tied development, including the rotation of the nucleotides with the formation of the double helix as a driving force, constitutes a "big bang" moment in the creation of life function. As it resolves the "chicken and egg" problem as to which one came first, whereby the answer is both and it was always tied since both were developed in an optimal intermolecular manner, is a satisfying answer. By having the form of the trigger and target derive from the same structure, and intertwine in the simultaneous development, an optimized structure with respect to both molecular configurations is achieved.

In developing this research, a striking result is the simplicity of the materials and in particular the process mechanisms involved in establishing the code of DNA nucleotide sequences. Considering that DNA developed soon after the formation of Earth according to present estimates, it had to be this way, that is simplistic and robust. There is not the need for a lengthy evolutionary process requiring multiple building blocks to synthesize and put into practice, as the materials and functionality of DNA are correlated. A confluence of unrelated probabilistic events is not required to generate molecular structures for lifeforms, or a sequential period of development of critical components: the development of the fundamental form of molecular life structure and function is achievable in a co-synthesized framework. Moreover, the development of the steroid molecules performed at the same time as the DNA nucleotides also simplifies the development of a critical life function as it assembles all molecules necessary within the same environment.

To produce the coded structure and alternative outcomes, the methodology indicated in this article, in which two cyclic structures are paired, and then paired again, can be utilized to form other structures that can contain encoded larger molecules as well as their molecular function enablers. This would be applicable to the formation of molecular storage devices, and the approach described in the article can be used to create alternative structures by manipulating the orientation of the molecules, and number of rings. Furthermore, the approach described in this paper can synthesize at the same time both the storage medium, as well as the access mechanism through the parallel development of the steroid triggering molecule. It is possible for precise replication of certain portions of the molecule through the use of the steroid triggering molecule, that is equivalent to transcription.

This methodology also should enable a questioning on how gene transcription may be performed. The new approach to transcription involves the direct connection with the steroid molecule to bind onto the DNA nucleotide, as it is pre-programmed since both were developed in parallel, and one influenced the other through intermolecular binding through co-structural development. Current strategy for transcription is through a receptor, hence indirect, but these results indicate that the steroid molecule is integrated with the DNA molecule, and is thus capable of inducing replication and transcription without the need for a receptor. These ideas that steroids and the DNA nucleotides share the same structure, and thus there is likelihood that the binding of the steroid hormone to DNA nucleotides triggers gene transcription as first proposed and developed in the preprint [21], and these results on the origin of DNA should solidify those ideas. The driving force for transcription is the stabilization energy provided by the steroid molecule that can be dissipated through destabilization about the bound area via lengthening of nearby hydrogen bonds along the nucleic acid strands. Thus, if the current methodology of protein induced transcription is correct, then there was a early life mechanism of transcription that did not require proteins, but rather utilized steroidal molecules.

The results indicate that the reason for purine and pyrimidine as the bases that comprise DNA as follows: Purine and pyrimidine together account for three of the four rings of the steroid molecule, and the coordination through hydrogen bonds accounts for the fourth ring. Yet another mystery that is further resolved by this work is the location of the missing hydrogen bond of the adenine-thymine pair relative to the cytosine-guanine pair. While A-T has two internal hydrogen bonds, there is an available oxygen element on thymine that can form an intermolecular hydrogen bond with the mid-molecule hydroxyl group of corticosteroids, such as the steroid hormone cortisol. Both results are further described in the preprints [21, 22]. This result, which is verified experimentally since prednisolone possesses activity whereas prednisone has no activity prior to conversion to prednisolone, provides hard evidence that the symmetry between steroid hormones and DNA nucleotides is more than coincidental, but is in fact functional. This result is important, as previously the A-T pairs were considered as "weak", but the complex can be made strong when coupled with an intermolecular hydrogen bond from corticosteroids.

A major advantage of this approach is that the translation of proteins from amino acids is immediately defined and consistent with the overall framework. Thus, the entire sequence of replication, transcription and translation is apparent, which is far further than alternative theories on the origins of DNA. The areas where additional development would include some of the details associated with the specificity of the amino acid to the functional sites of the steroid molecule. In addition, the amino acid sequence on the other side of the strand, which does not survive the rotation into the double helix is also interesting to evaluate. This will aid in the determination of the initial set of amino acids that were used to translate into the first encoded proteins.

The steroid molecule apparently is an inherent structure of DNA, as it is difficult to argue against that there is an embedded structure traced into each and every nucleotide pair comprising nucleic acids. Thus, these results have defined the structural basis of DNA as the steroid molecule. There are a variety of similar structures within that family of molecules that can be used as the trigger for DNA to enable transcription. Thus, steroid molecules should undergo close examination for health benefits than they currently are. Corticosteroids are a first line of defense in inflammatory diseases, but other forms of steroids, including hormone replacement therapies, and other forms of synthetic steroid hormones should be closely studied, as currently there seems to be a barrier attached to these molecules. Life function is inherently derived from the steroid molecule, and thus its use for therapeutic purposes should be closely assessed.

It is also noted that the integration of the steroid molecule in this construct can also be extended to the concept of using the molecular structure to match-up single-strands of pre-synthesized RNA. Thus, the steroid structure could organize along the lines of the base pairing, as it was associated with a base four structure. Thereby, the steroid molecule could serve as a template using additional functionality along the interior to align the amino acids. Thus, the procedure described could be adapted into an environment dominated by RNA molecules, and how these structures can be extended to derive encoded proteins.

The structural symmetry of steroid hormones and DNA nucleotides is a practical result, for example in the area of design of therapeutics. It was indicated that each functional element of the steroid hormone can be assigned a purpose, such as the mid-molecule OH group to form an intermolecular hydrogen bond with thymine, the end groups to form an ionic coupling with the phosphodiester backbone, and the internal structures, such as rings, methyl groups, and double bonds to enhance stability and interaction. Furthermore, the backside elements of the steroid hormones, such as an OH group, can influence the transcription process. Indeed, these concepts have been applied to RNA based applications in forming an intermolecular bond with uracil in a like manner of thymine [25].

As the interaction of binding the steroid molecule onto the strands of the DNA to initiate strand separation, which would thereby enable replication and transcription, a type of molecular machine was synthesized. It can be envisioned that the DNA strands are in fluidic motion, and the pinching induced by the steroid elements, enabled a bubble to be generated on either side of the interactive zone, which would permit access by chemical replicating agents. Depending upon the type of environment, the steroid triggering agents can be evolved to permit directed or sustained replication periods. The transition to ionic bonding interaction of the phosphodiester chain of the steroid triggering molecule is an example of the need for sustained interaction with the DNA strands. It is noted that the approach is capable of working with ionized or unionized forms of the phosphodiester backbone.

As presented by the checklist in the introductory section regarding the requirements for a valid new approach to explain the creation of life function, the approach developed in this article satisfies each of the requirements. It is a truly remarkable that DNA nucleotides and steroid molecules were developed as a unified complex, developed in an intermolecular manner with each molecule influencing one another to enable a big bang, instantaneous moment of the simultaneous synthesis of structure and function. Thus, as soon as the steroid molecules are ejected from the DNA-steroid-amino acid complex concurrent with the DNA nucleotides rotating into a position consistent with the double helix, the result is instantaneous life function.

Chapter 11

Summary

The origin of life is one of the greatest mysteries. The mechanism for the synthesis of DNA is synonymous with the chemical origin of life, and theories have been developed along many lines of reasoning, but resolving all requirements remains a challenge, such as defining an objective path to produce sequences of encoded nucleotides paired as adenine to thymine and guanine to cytosine. Here, a new solution to quantify the origin of DNA is presented. The approach is based upon three lines of experimental evidence and agreement of structural symmetry between DNA nucleotides and steroid hormones, and introduces a new concept of synthesizing both structural and functional characteristics of DNA at the same time within a single unified complex of interleaved steroid structures which form reaction vessels that serve as co-enzymatic building blocks. The new approach indicates that the establishment of the DNA nucleotide "code of life" is among the very first synthesis steps. Moreover, as a consequence of the intermolecular synthesis of both structural and functional characteristics within a unified complex, there is a culminating process step that sets forth in motion both DNA and the steroid molecules resulting in the instantaneous release of life function.

A complete solution to the communications and information generation of DNA was shown, including its encoding, transmission, decoding, and translation. Encoding of DNA was shown to comprise a co-enzymatic complex in association with steroid molecules, in which the relative orientation of a ketone group to induce intermolecular coupling in paired construct of steroid molecules. The result was a base 4 nucleotide code sequence that is brought to function by a set of two types of steroid molecules, of which the collective of DNA and steroid enables both structure and function, and therefore the instantaneous creation of life function when released from the complex. Transmitting DNA was shown through the double helix, which served a purpose in releasing the steroid molecules, which can later join with the DNA complex to induce transcription. Decoding was performed directly on the strand by associating the steroid molecule in paired alignment with the nucleotide base. This provides a uniform surface from which to form an interaction vessel to assemble amino acids, which are transferred to the structure by the steroid molecules, which together forms a ceiling of the reaction vessel. Upon stabilization, the amino acids form a peptide bond and a protein structure. Error correcting is performed in conjunction with the steroid molecule, which was co-developed with the DNA nucleotides within the same complex, and thus provides the integrity of the information content contained within the nucleotides. As can be seen, although DNA gets all the glory, steroid molecules play a major role not only in forming the structure of DNA, but also in its initial function in producing proteins, and are thus of equivalent importance to achieve life function.

Chapter 12

IP-Protected Product Directions

It is correct that discovering an origin for DNA will not only explain the past but also provide a primer for the future. In this regard, there are innovations that derive directly from the basic discovery of the structural symmetry and functional purpose between steroid molecules and DNA. Work by the author has already begun in analyzing mechanisms of action and novel drugs that can interact directly by binding to nucleic acids, either single-strand in a secondary structure or as a double strand [25]. Other aspects of this work involve technology and philosophical advancements, and thus there is a wide array of topics that will be influenced by these results, assuming that a team is effective in delivering the message. Fortunately, the results were anticipated, and several provisional patents were filed along the way of the development path, such that it will be possible to bring in external investment and develop the science and technology in a coordinated way.

12.1 Medical, health, and biological

From these results, new advancements in the medical field will result, and medical theories will need to be reexamined.

- New drugs and supplements based on intermolecular binding directly to the available functional group of uracil in RNA-based diseases or thymine to enhance physiological control
- Synthesis of new life forms based on tetra-ring formulation
- New life functions based on different ring structures and hydrogen bonding configurations

- Systems and cellular models integrated with nuclear interaction with steroid molecules for predictive modeling of basic core functions such as body temperature control

- New understanding of transcription mechanisms and the interaction of intracellular receptors with steroid hormones and DNA

The developments can lead to new results on intermolecular binding to RNA viruses. For example, the abstract of [25] follows: Exogenous RNA comprises the genetic material associated with several diseases which require immediate treatment, and thus mechanisms to hinder intracellular translation and reproduction of RNA viral agents are of great importance. Applying recent developments from this lab in methods relating to the interaction of DNA with steroid hormones, cyclic compounds are presented for intermolecular binding to nucleic acids. The requirements to achieve binding with RNA nucleotide pairs are described, which involve at a minimum functional elements positioned to interact with the lateral phosphate groups for each of the RNA strands through coupling with a positively charged ion, such as Mg^{2+}, Ca^{2+}, or Zn^{2+} ions; and an intermolecular hydrogen bond with the oxygen element of uracil at the carbon two location. Additional features of the binding molecules are examined for enhancements and differentiation in binding capability and include aromatic groups that have both a structural role of steric hindrance and a functional role to stabilize the binding mechanisms. Several categories of cyclic compounds are associated to have specific binding capabilities, and the interaction of these structures with potential receptor molecules are evaluated for assessment in delivery and binding of the compound to nucleic acids.

12.2 Technology

The advancements described on the design of DNA and genetic codes can enhance technology endeavors, for example:

- Artificial Intelligence coupling with molecular strategies

- Data storage techniques with internal read function

- Adaptive material synthesis that conform simultaneous structure and function

- Molecular machines with inherent correction of malfunction

- Design of computational engines of molecular configuration

12.3 Philosophical

Even at the basic thought levels in examining the past and future, the results on the design of DNA will enable a scientific basis for discussion as it has aspects of agreement with the central beliefs of several philosophies:

- Search for extraterrestrial life and seeding other planets for life, which will be of interest to national organizations, such as NASA, as materials such as chrysene are relevant.
- Evolutionary theorists will need to put the new results into context including stochastic analysis to answer fundamental questions, which will be of interest to a range of funding organizations and academic institutions.
- Placing the results within the context of theological tenets, which would impact the general population to contemplate the meaning of the process of encoding DNA by pairing pairs, and through the formation of the double helix while releasing steroid molecules to enable function thereby resultant in the instantaneous creation of a life function. We were created in the image of a master pattern, which was a one-time use. The double helix is a lower information content than the original master, which contained structure and orientation, and molecular functional compounds.
- A basic need for life balance is evident: DNA denotes the intellect, the mental capacity. Steroid denotes action, strength, the physical capacity. Both were derived in approximately equal amounts from the master pattern. Thus, there is an inherent need to strive for both mental and physical performance. Doing only one and forgetting about the other is not in accordance with the master design.

Appendices

Appendix A

Mechanisms

Reaction mechanisms are indicated to form the differentiated molecular structures of: (a) adenine-thymine; (b) guanine-cytosine; (c) steroid type paired with A-T; (d) steroid type paired with G-C. The pairing of steroid molecules and DNA nucleotides works both ways, which means that the orientation of the steroid molecule impacts the synthesis of the DNA nucleotides, and vice versa, since the interaction of two results in blocking elements which prevent amination or reduction from occurring. This section is presented with an indication of the synthesis steps that result in differentiating function, including the separation into the base pairs. Then the suggested reaction steps for producing the final products is indicated, which includes the steroid molecules. Finally, the common steps are described, including the possible starting molecules from which the steroid molecules results. There are other synthesis pathways than the ones suggested in this section, however, the overall logical sequence would be similar in terms of the transformation of the molecules into the final products. It is important to note that the synthesis steps occur within the same environment, such that while one base pairing is being synthesized, another aspect may be taking place in parallel, and thus the overall reactive environment needs to be consistent across most steps. At prebiotic conditions and environments, UV illumination will play a significant role and catalyze reactions for amination to take place at appreciable rates via free radical, photochemical reactions [26], as implied in the developments of this section. In addition, as noted in Section A.1, the reactions take place within three dimensional constructs, although some mechanisms are presented in a standard two dimensional framework.

A.1 Co-enzymatic Coupling

The discovery that I made of a primer molecular shape embedded within the structure of the complementary bases comprising each DNA nucleotide pairing as described in Chapter 2 drives the development of the co-enzymatic analytic framework described herein. The common structure indicates that there is a correlation of the structure between the DNA nucleotides and steroid molecules as an enzymatic construct of interleaved tetra-ringed structures, of which one section is covalently bonded to a phosphodiester backbone, while the adjacent structure is connected through hydrogen bonding in close proximity. The result is an arrangement of reaction vessels and co-enzymatic framework for the co-development of sequences of encoded structures, as defined through intermolecular hydrogen bonding, and the relative orientation of the tetra-ringed structures.

As indicated in Figure A.1, the coordination of the paired building blocks enables a sequence of effective reaction vessels, secured at the ends through phosphodiester covalent coupling. This confined environment enables the potential for an enzymatic capability in which the temporary formation of bonding arrangements to permit intermolecular cyclic structures that can establish element substitution, the degradation of chemical bonds to reduce intramolecular stress, and the differentiation of the templates into their individual capacity as a nucleotide base. While the structures have aromatic stabilization, the intermolecular structures benefit from resonance stabilization, and thus the intermolecular constraints to maintain an in-plane structure during the synthesis steps promote the present-day configuration in which the original aromatic structure was traded for an amine based construct of a planar configuration.

A.1.1 Chain linkage

In Figure A.2, the method of linking stacks of molecules together is indicated, in which the enzymatic characteristics of the implicit reaction vessel are implemented. The initiating step involves the amination of what will be the steroid molecules, that is the non-nucleotide molecule. This occurs at both ends of the molecule, in proximity to the diketone group on one side and the hydroxyl group on the other side, as indicated in Figure A.2(a). Because this planar structure will enable a catalytic activity, the amine group reacts with both of the ketone groups, and the amine group achieves alkene addition on either side of the phenol group, as shown in Figure A.2(b), which then integrates within the DNA nucleotide molecule, displacing the residual as methane and as methanol. This reaction is similiar to the azofullerene synthesis [27,

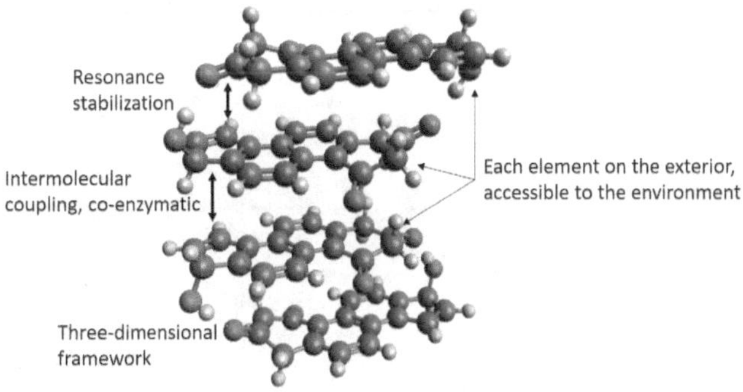

Figure A.1: By interleaving the tetra-ringed aromatic structures, and securing each alternating pair to a phosphodiester backbone, there exists the capacity for an enzymatic function in the development of each molecule. This framework constitutes a series of reaction vessels, which together with aggressive environmental conditions, the co-synthesis of the DNA nucleotides, and its triggering elements, comprised as steroid molecules, result from interactive co-development.

28, 29]. The connection is then achieved by binding onto the inserted amine groups, and then the formation of the sugar group to stabilize the phosphodiester backbone, as indicated in Figures A.2(c) and A.2(d), respectively. Relevant reaction sequences to this synthesis route are the stereoselective Diels-Alder reactions [30, 31, 32], for this type of integrated structure.

This procedure of coupling one of the molecules for each pair results in a construct of a chain of alternating pairings of tetra-ringed structures, secured on both sides by a phosphodiester sugar backbone. The non-covalently bonded molecule is secured through hydrogen bonding associated with the PO_4H molecule on either side. The spacing of roughly 7 Å between each pair of couplings is consistent with the aromatic stabilization of both molecules and the physical distance of the phosphodiester backbone of DNA. This consistency is thus further experimental proof correlated with these developments.

A.1.2 Differentiation

The implementation of this configuration with positioned oxygen elements involves intermolecular hydrogen bonding to cross-couple the two pairs, which is critical in defining the differentiation between A-T and C-G pairing. Since the ends are already coupled to connect the pairs, the available oxygen structure of the di-ketone group is arranged through the middle of the molecule, that is the second or third ring. The

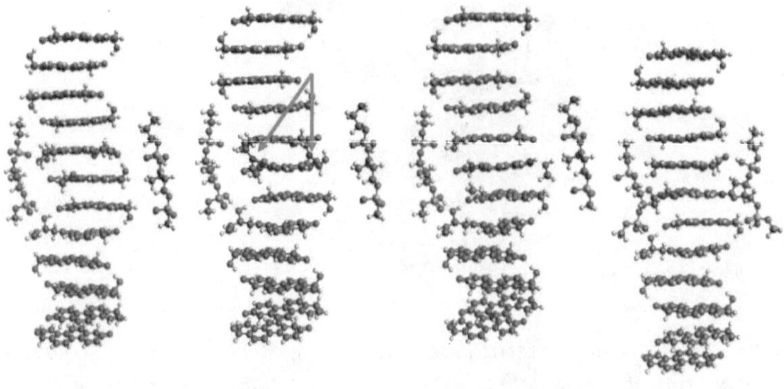

(a) Amination (b) Connect (c) Separate (d) Bond

Figure A.2: (a) To initiate the link of the phosphodiester backbone to the DNA nucleotides, the two amine groups are attached to the molecule, which will become the steroid molecule. (b) a binding is achieved to replace the alpha carbon between the diketone groups, and on the other end, the carbon binding to the hydroxyl group is replaced. (c) after binding, there is tendency for leveling with respect to the steroid molecules above and below the molecule where the amine group was replaced. (d) the final orientation, and ready for further amination and separation.

ketone group is configured two carbons away from the ketone group associated with the end-based pairing. This ketone group is then connected to the ring, or via a hydroxyl group, connected in proximity to the ketone group on the other molecule. The position of the ketone group relative to the adjacent steroid molecule of the adjacent pair will define whether the molecule becomes A-T or G-C. Interestingly, the ketone group associated with the A-T pairing of the steroid molecule will cross-couple to the nucleotide molecule and reduce the chances of amination from occurring, and thus arrive at an adenine structure. This is indicated in Figure A.3(a) and (b) for G-C and A-T respectively, indicating the position. This is demonstrated by looking at the electrostatic potential in Figure A.3(c) and (d). This provides the underlying basis for the differences between A-T and G-C. Here an OH group on the receiving molecule is included, but simply the presence of two oxygen elements on both sides of the molecule, one oxygen affiliated with the hydrogen bonded steroid molecule and the other oxygen associated with the covalent connected molecule, with respect to the phosphodiester backbone.

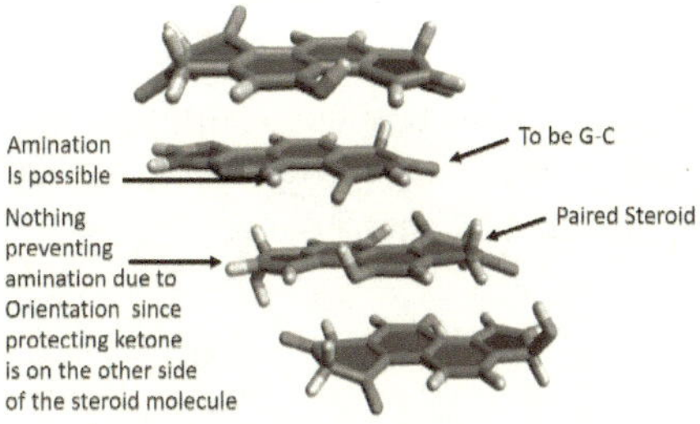

(a) G-C

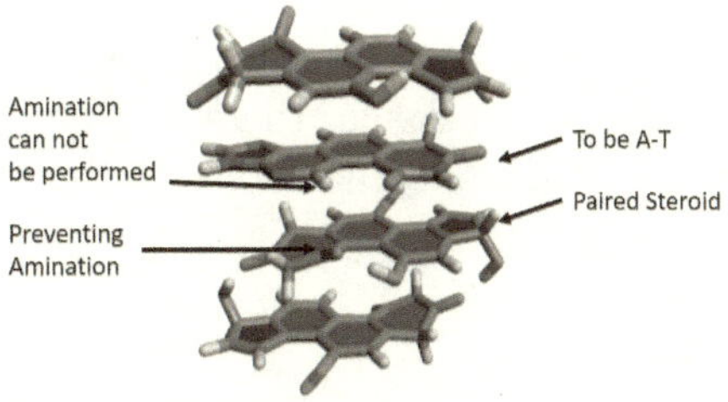

(b) A-T

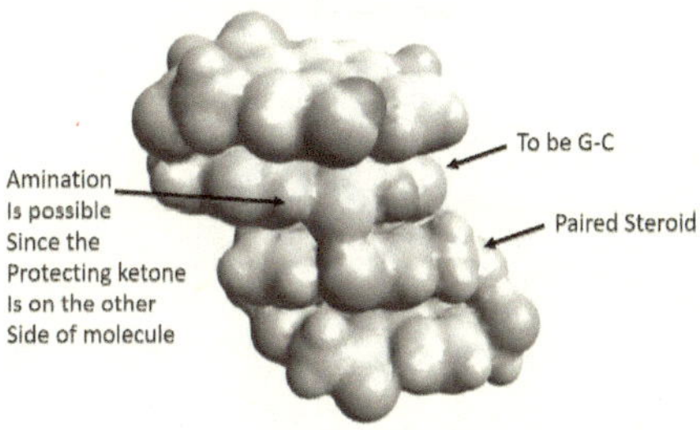

(c) G-C

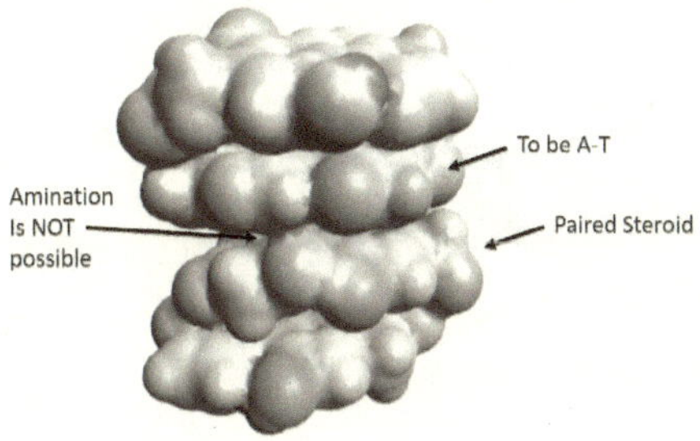

(d) A-T

Figure A.3: (a) For G-C, the orientation of the steroid molecule, second from the bottom, with respect to the molecule that will become a G-C nucleotide is oriented such that amination is possible on the leading element of guanine, as initiated through UV illumination. Thus, an NH_2 group will be affiliated with the leading side of guanine. (b) For A-T, however, there is a connecting group of O-OH associated with the steroid molecule pairing, thus amination is not possible. Ultimately, this is why A-T has two internal hydrogen bonds, while G-C has three. (c) The electrostatic potential indicates that the steroid molecule will not influence the G-C amination. (d) The electrostatic potential indicates the coupling with the steroid molecule which will prevent amination for A-T.

In addition, this issue is also seen on the other side of the molecules, where the access of the molecules, and their relationships between the paired relationship of the steroid molecule to the nucleotide molecule, which will ultimately become the DNA nucleotide base pairing. In Figure A.4(a) and (b), the other side of the steroid - DNA complex is indicated. The electrostatic potential indicates the interaction of the steroid agents and the molecule that will become the DNA nucleotides in Figure

A.4(c) and (d) for the G-C and A-T pairings. In addition, the pairings for C-G and A-T also show similar results, both on the front side and the back side. On the A-T backside, there are no oxygen elements associated with the di-ketone grouping, while on the C-G backside, there will be one ketone group associated with the diketone group. This differentiation will enable a differentiation in the subsequent reaction sequences.

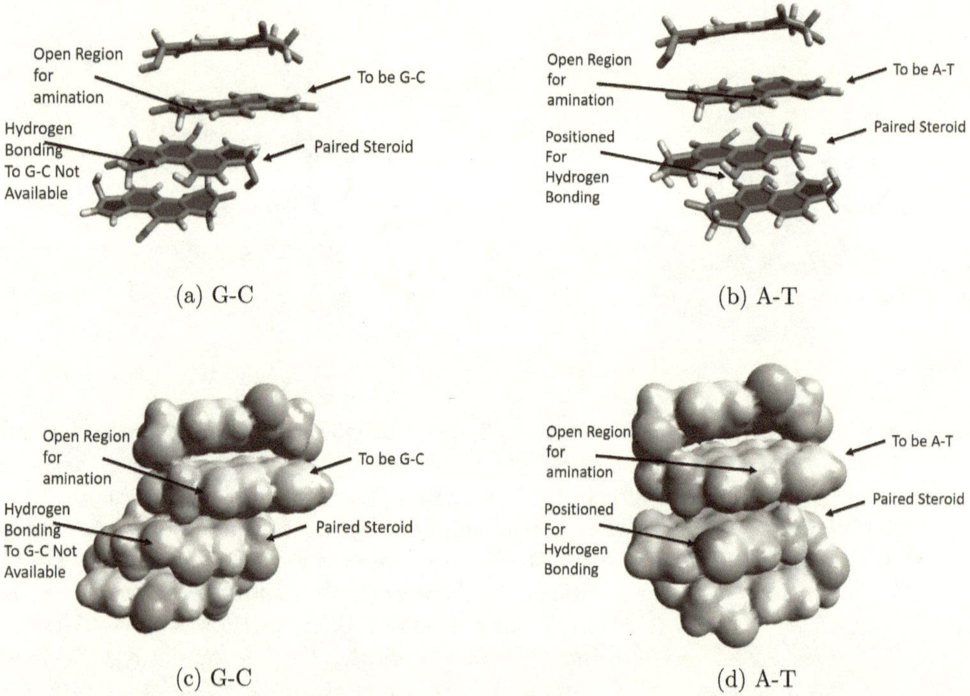

(a) G-C (b) A-T

(c) G-C (d) A-T

Figure A.4: On the other side of G-C relative to Figure ??(a), the orientation of the steroid molecule, second from the bottom, with respect to the molecule that will become a G-C nucleotide indicates potential influence over the oxidation and amination. (b) For A-T, there is little association with the backside other than a potential hydrogen bond, and thus oxidation will be effective at the thymine molecule side. (c) The electrostatic potential indicates that the steroid molecule will influence the G-C association on the backside. (d) The electrostatic potential indicates the coupling with the steroid molecule will influence oxidation for A-T.

In binding the phosphodiester backbone prior to its disassociation, as shown in Figure A.5, of which the binding occurs at both ends of the nucleotide molecules 1, 3, 5, 7 and 9, while hydrogen bonding stabilizes the steroid molecules 2, 4, 6 and 8. Note that phosphor group contains both OH and an O, such that hydrogen bonding can be achieved at both ends of the steroid molecule: one side of the steroid molecule contains an OH group which will hydrogen bond with the O element of the phosphate group, and on the other side, the steroid molecules comprises a ketone group which will hydrogen bond to the hydroxyl group of the phosphate group. The conversion of the link to a nitrogen element is indicated, and the binding occurs either at the middle carbon located between two ketone groups of the six-membered ring, which is discussed in the subsequent sections. It is noted that there is sufficient space for occupation of the steroid structure in between each DNA nucleotide, as both structures are relatively planar.

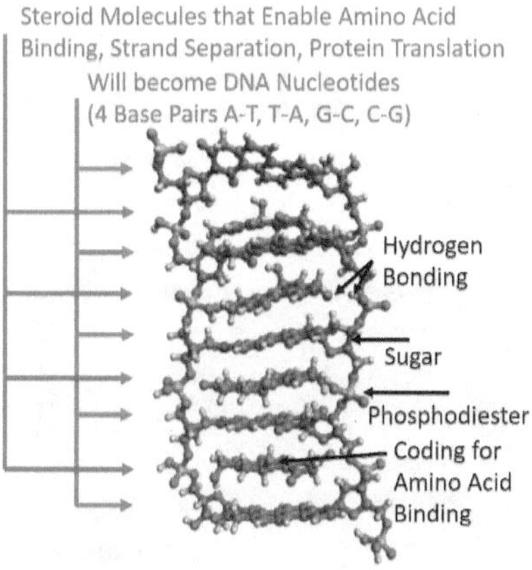

Figure A.5: Model indicates the paired synthesis of the steroid molecules and the DNA nucleotides during a stage of fabrication where the DNA nucleotides are formed, but not yet separated into their basic base pairing with associated hydrogen bonding. There is interaction with the steroid molecule during the formative stages, and the DNA nucleotides influence the formation of the steroid molecule, and the steroid molecule influences the DNA nucleotides. The association with amino acids is not included here.

Continued amination, oxidation, and reduction steps are performed to complete the co-synthesized fabrication of not only the encoded DNA nucleotides, but also the steroid molecules. These steps are described in the following section. Interestingly, at the end of the synthesis steps, the steroids are released from the unified complex when either the purine of pyrimidine molecules that are out of position rotate into position, that is with the purine molecules back and pyrimidine molecules forward on the 5' side, to form the double helix. Thus, there is a culminating step to the synthesis which brings both molecules to life, as the steroid molecules can interact with the DNA double helix to enable replication and transcription.

The overall procedure to produce a differentiated structure is indicated in Figure A.6 for the adenine-thymine pairing. In Figure A.6(a), the differentiation and separation of the DNA nucleotides is seeded by the amination step onto the alpha hydroxy of the steroid molecule, with subsequent attack to the DNA nucleotide in a position to aminate the rings of the purine molecules. Figure A.6(b), an amination at the steroid and a coupling to the C ring of the DNA nucleotide, when still intact is achieved to initiate the separation. Figure A.6(c) completion of the amination, and decoupling of the B ring with oxidation onto thymine completes the process. Figure A.6(d), with the removal of the ethene molecules, there is self alignment of the pairs to complete the transformation.

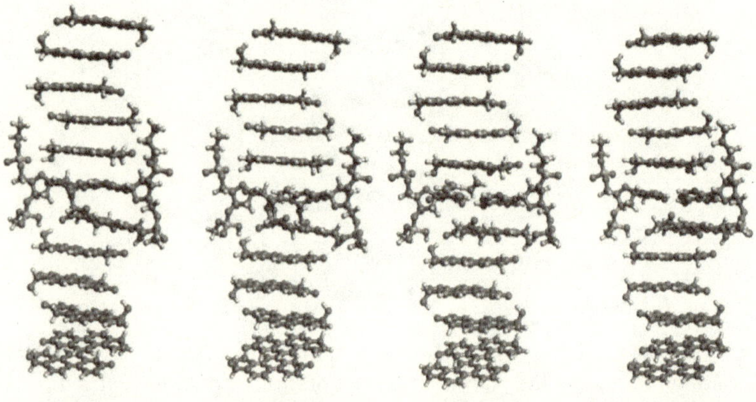

(a) Initial (b) Bonding (c) Aligning (d) Final

Figure A.6: (a) The differentiationand separation of the DNA nucleotides is seeded by the amination onto the alpha hydroxy of the steroid molecule, with subsequent attack to the DNA nucleotide in a position to aminate the rings of the purine molecules. (b) An amination at the steroid and a coupling to the C ring of the DNA nucleotide, when still intact is achieved to initiate the separation. (c) Completion of the amination, and decoupling of the B ring with oxidation onto thymine completes the process. (d) with the removal of the ethene molecules, there is self alignment of the pairs to complete the transformation.

A.1.3 Separation

As indicated in Figure A.7(a), the reaction mechanism suggested for the formation of adenine-thymine is indicated. It is important to note that this reaction takes place in conjunction with pairing to the steroid molecule, which has hydrogen bonding to the front portion of the adenine-thymine group, which prevents amination from occurring on the adenine front side. This provides a key differentiating step. In addition on the backside, the interaction leads to the preference for oxidation on the thymine side. Aminations in an ammonia environment, with availability to oxidation components is performed. The entire structure is under stress, in which there is a significant driving force to form the double helix, and thus enable the separation of the base pairs. For the G-C base formation, the suggested reaction mechanism and strategy is indicated in Figure A.7(b), which occurs within the same environment as that of A-T. Recall that because of the paired steroid molecule positioning of its ketone elements, it is possible to aminate the front of the molecule, and thus there is amination on guanine. This enables the oxidation at the thymine side. As with A-T, the separation is driven by the drive to form the double helix, and its lower energy state.

Thus, this provides the basic answer as to why the A-T pairing only has two internal hydrogen bonds, while the G-C pairing has three internal bonds: because of the relative orientation of it to the paired steroid molecule permits hydrogen bonding. Moreover, it indicates why the overall structure of the base pairings occurs, and provides the driving force for it to occur, high stress, and a significant reduction in energy by transitioning to a double helix form. It will also provide answers to why $NH2_2$ is on the pyrimidine purine side for G-C but on the purine side for A-T, and other such comparative matters. It is important to note that these reactions are effectively taking place over the entire molecule, so the same process steps have to be addressed concurrently on how they would react to A-T if working on C-G, for example.

A.1.4 Completion

The final steps to produce the final products are all performed in the same reducing and methylation environment. In Figure A.8 the steps to reduce the oxygen group on thymine is indicated followed by the methylation to form the methyl group associated with thymine. Since the oxygen group of thymine is not engaged in hydrogen bonding and has an adjacent available carbon segment, this will allow for its reduction. In Figure A.8(b) the reaction sequence is indicated to reduce the oxygen functional group of what will become cytosine. The positioning of the oxygen group enable its

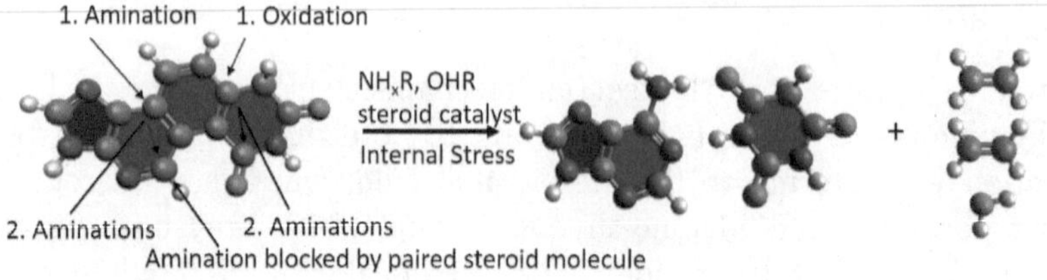

(a) Formation of A-T

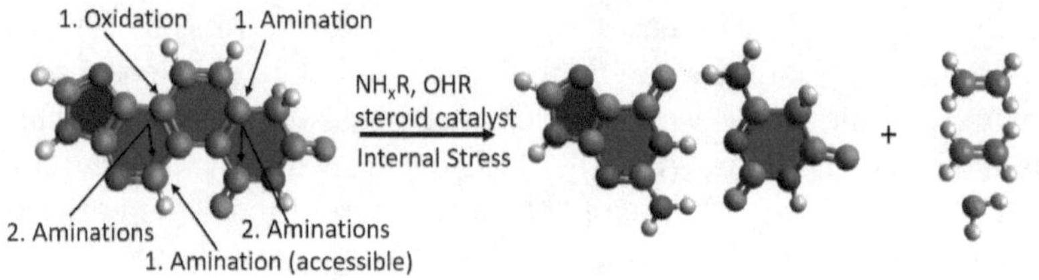

(b) Formation of G-C

Figure A.7: *(a) For the A-T base formation, it is important to recall the result that indicated because of the paired steroid molecule positioning of its ketone elements, it is not possible the aminate the front of the molecule, and the amination occurs on the purine side, adenine. This enables the oxidation at the thymine side. The separation then occurs though an amination on both sides of the molecule, with the double bond established on adenine. (b) For the G-C base formation, it is also important to recall the result that indicated because of the paired steroid molecule positioning of its ketone elements, it is possible to aminate the front of the molecule, and the amination occurs on the purine side, guanine. This orientation also enables the oxidation at the guanine side, and hence amination at the cytosine side. The separation then occurs though an amination on both sides of the molecule, with the double bond established on cytosine.*

reduction, whereas the other segments are engaged with the guanine base, and thus do not undergo reduction.

Finally, and in parallel, in Figure A.9(a) the rearrangement of the steroid in a corticosteroid function is indicated, in which the planar configuration is sought since intermolecular hydrogen bonding with the adjacent molecule is not required due to the covalent coupling with the phosphodiester backbone, and hydrogen bonding to the phosphate molecule. Likewise, the intermolecular coupling with the C-G pairing through hydrogen bonding is not available due to the positioning of the oxygen

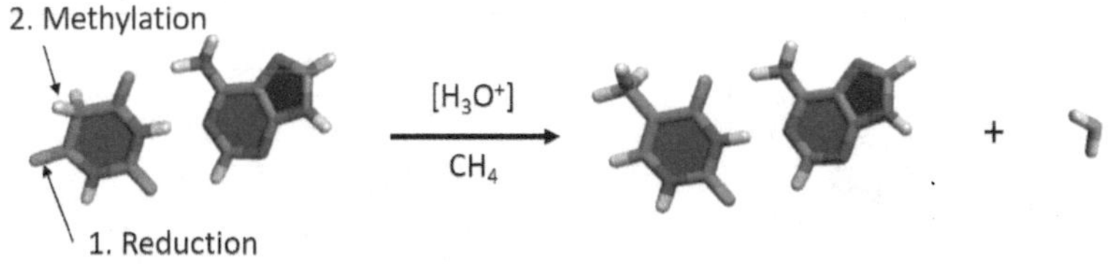

(a) Formation of Thymine

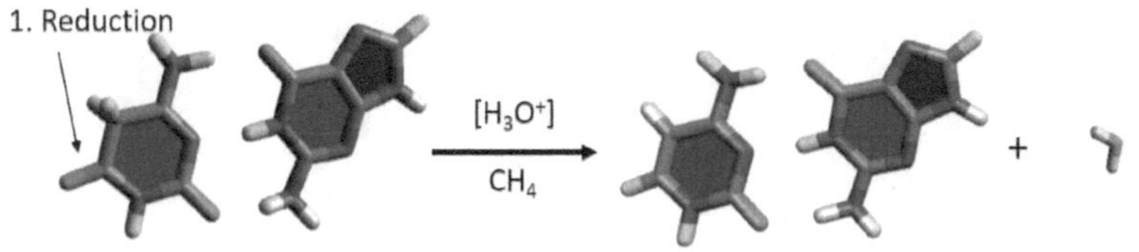

(b) Formation of Cytosine

Figure A.8: (a) Steps to reduce the oxygen group on thymine and the methylation to form the methyl group associated with thymine. The oxygen group of thymine is not engaged in hydrogen and has an adjacent available carbon segment which will allow for its reduction. (b) The reaction sequence reducing the oxygen functional group of cytosine. The positioning of the oxygen group will enable its reduction, whereas the other segments are engaged with the guanine base.

group, and thus will undergo elimination. The result is the steroid structure, planar configuration, as indicated in Figure A.9(b). These molecules will then be useful for further usage in constructing the interaction vessels.

A.1.5 Starting materials

To broaden the scope, the developments will consider a related structured, the fused aromatic four ring structure of chrysene, indicated in Figure A.10(a), which is a relatively simple organic structure. From a structural perspective, the molecule can be aligned with itself with more precision in comparison to a linear configuration. It is also a stable compact structure, and further is the same general structure of a steroid molecule, and is also consistent with the general structure of both the DNA nucleotide pairs and the steroid hormones. With the addition of the oxygen groups to the ends of chrysene, stacking of the molecules can be secured through attraction at the common alpha position situated in between the two ketone molecules. Further, this molecule is active and can be reduced as indicated Figure A.10(b), which would be enabled

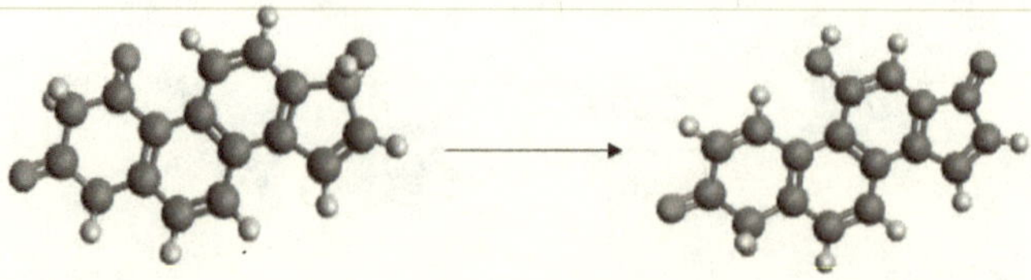

(a) Steroid for A-T pairing

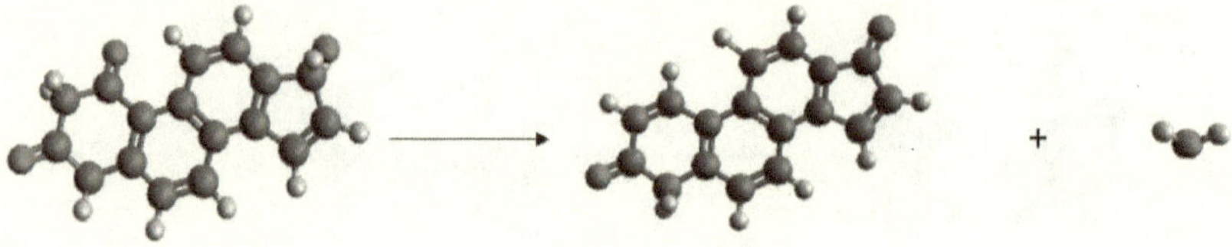

(b) Steroid for G-C pairing

Figure A.9: (a) For the steroid molecule paired with A-T, the planar configuration is indicated in which the position of the oxygen is paired through hydrogen bonding with the A-T pairing. The ends will remain intact since there is hydrogen bonding interaction with the phosphodiester backbone. (b) A similar result is obtained for the steroid molecule paired with C-G, except that because the mid-molecule oxygen groups are not engaged with C-G, they will be reduced. Hence, there are two types of steroid molecules produced.

through a catalyst, as with all the reactions, and incorporates photochemical effects.

Chrysene is a relatively common material, especially in astrobiology analysis, as it is among the class of polycyclic aromatic hydrocarbons. As non-polar molecules, polycyclic aromatic hydrocarbons are located in the thermal decomposition of organic materials and are found in tar and carbon deposits. The molecules are commonly found within the universe and are often associated with the formation of new stars. They may account for a significant percentage of carbon within the universe. To enable this conversion, chrysene will follow Clar's rule which indicates the positioning of the Kekule resonance structures with the largest number of disjoint aromatic n-sextets for polycyclic aromatic hydrocarbons, as presented in Figure A.11.

This will enable a difference in structures at the ends of the molecule, which are relatively easy to attack the projecting hydrogens, thus inducing the structures of Figure A.10(b). Thereby, it will enable a hydrophilic sections at the ends, and hydrophobic in the middle of the molecule. This arrangement will maintain the paired structure with hydrogen bonding at the ends and molecular similarity in the

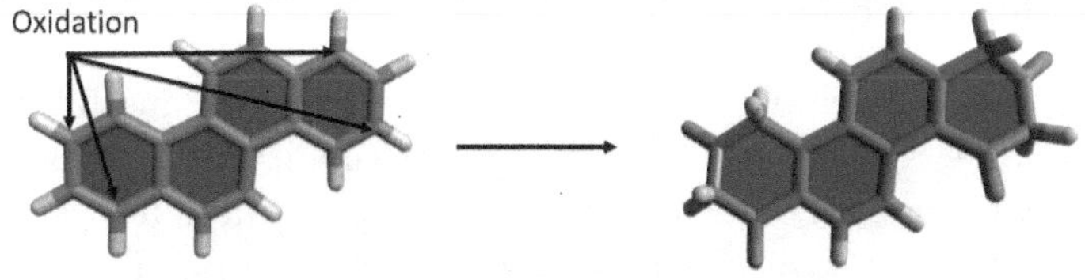

(a) Conversion of chrysene

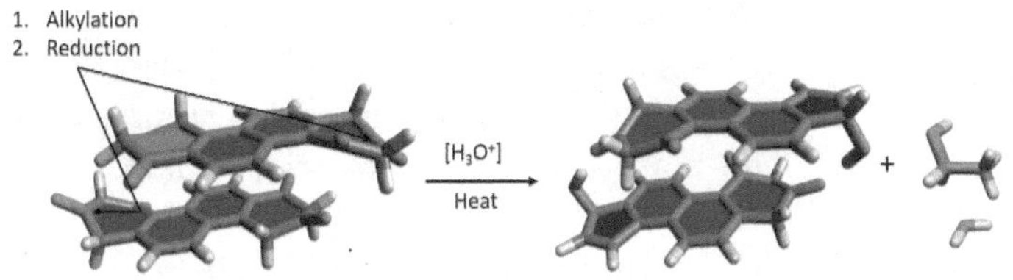

(b) Steroid complex

Figure A.10: (a) Conversion of chrysene to a diketone on both sides of the molecule. (b) The reaction to reduce the chrysene based structure to the steroid structure is indicated.

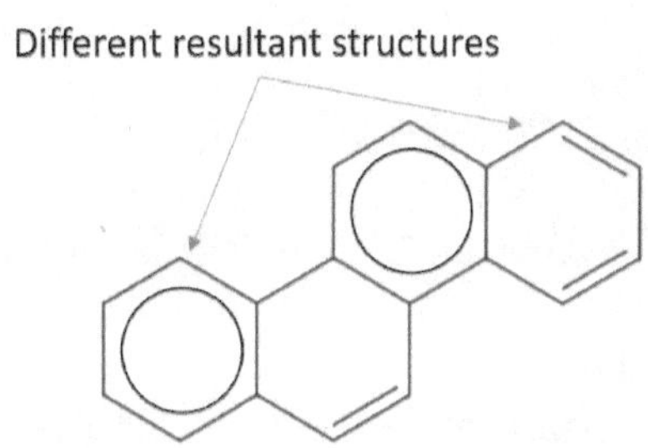

Figure A.11: Aromatic structures of chrysene according to Clar's rule, will enable different structures on the ends, inducing hydrogen bonding in pairs, and alignment in the central portion, inducing hydrophobic similarity.

central area, and thus act as a filtering mechanism as to the association of higher concentrations of like molecular pairs.

A.1.6 Alternative sequences

The general strategies developed in this section can be slightly modified, but the general plan is the same. For example, there is the possibility that the steroid molecule started out as six membered on all four rings, with the reduction to a penta group on the DNA nucleotide done while the attachment to the phosphodiester backbone was taking place. Other arrangements in the hydrogen bonding configuration are possible, including the possibility of additional functional groups along the periphery. However, the basic scheme remains unchanged.

Other sequences, which are similar in scope, can be developed with hydrogen bonding obtained for the middle section of the molecule through projections along the hydrophobic edges. The main concept would still remain thought, which is that the orientation of the rings is the overriding and exclusive factor defining the differentiated capacity of the intermolecular pairings to result in either an A-T or G-C base configuration.

A.2 DNA strand separation

While the DNA nucleotides are now in a stable double helix configuration, in order for it to have purpose, it needs to be accessible for reading and action. It is good that the double helix is so stable, but it also limits its availability for opening and reading, and without that capability it is just an interesting polymeric molecule, but would serve little purpose. Fortunately, it was co-synthesized with its key, the steroid molecule, that is capable of opening it. Therefore, in a transmission format, it is necessary to transmit not only the DNA molecule but also its key to opening it. Therefore, it is favorable for DNA, and its paired molecules, to be coupled in a cell format to contain the molecules.

Thus having produced the paired and encoded structures of both DNA and steroid molecules, the bases are accessible for replication and transcription or translation by binding from the steroid molecule onto the double helix to enable strand separation. This occurs through stabilization of the strands, which are vibrating and oscillating molecular constructs of DNA. DNA is known to have natural modes of oscillation [33]. There is kinetic energy, which comes from the vibratory nature of the DNA strands, and thus is transferred to potential energy through the stabilization process from the binding of the steroid molecule. The overall energy function must be balanced and thus the external reduction of internal energy must be dissipated by providing that energy back to the external. This can happen through molecular disorganization at neighboring sties and hydrogen bond breaking in this case. Approximate mathematical representation for the process is indicated in Equations A.1 and A.2, for the energy dissipation, $E_s(x)$ about the binding site, and the strand separation from its nominal position, $S(x)$. The equations characterizing the strand separation are given by:

$$E_s(x) = \frac{1}{2}(1 + \tanh(k_x x)) \quad (A.1)$$

$$S(x) = k_b(1 - (\tanh(k_x x)^2)) + S_n \quad (A.2)$$

where the parameters are associated with hydrogen bond energy $k_x = 1/40$, coupling k_b=3, and nominal separation S_n=1.8 Å, and x runs from 0 to 200 nucleotides, with the result than mirrored from -200 to 0 nucleotides about the binding point. These equations can be used to approximate the strand opening away from the binding site. It is dependent upon the energy associated with the hydrogen bonding keeping the strands intact, as when the term $\tanh(k_x x)$ approaches unity, the strand separation will be equal to its nominal value. Other formulations for strand separation can use wave equations to describe the motion of the DNA strands, which will be constrained by the cross-coupling of the steroid molecule, and thus result in reflected wave patterns that will induce interference patterns, and bond separation for bubble formation.

As the DNA molecule is stabilized by binding with the steroid molecule, there must be a balance in energy in terms of a release through the disorganization of the non-covalent weakest link, that is the breaking of hydrogen bonds such that the overall energy is nearly equivalent to that of the stabilization provided by the binding event. Essentially, the DNA strands act as two strings in constant motion, which is fluidic, and with the cross-bonding induced by the steroid molecule, that is the triggering element, it acts as a pinch-point, and the strands, which were in relatively independent motion become constrained at one point, and around that point, a small bubble must form through hydrogen bond breaking upstream and downstream of the pinch-point. In Figure A.12(a) the binding of the steroid molecule onto the formed DNA complex is indicated in a molecular model as to its feasibility. In Figure A.12(b) the electrostatic potential is indicated of the steroid molecule to show that the negative potential of the phosphodiester zone will enable alignment of the steroid molecule for binding. In Figure A.12(c), the profile of the strand separation is approximated on either side of the binding site.

It is possible to deploy multiple steroid molecules to induce a stronger association and knoting of the DNA strands, and thereby improve the duration of the subsequent opening of the strands of the nucleic acid. Thus, with intermolecular binding the strongest, it would favor chains of intermolecular coupling TA to form stabilization

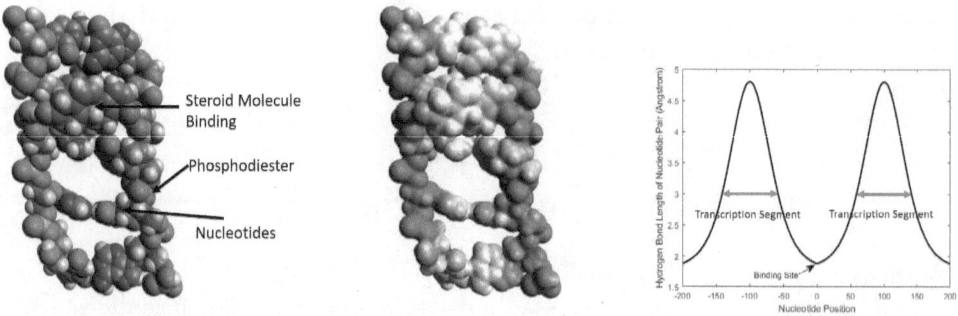

(a) van der Waals spheres (b) Electrostatic Potential (c) Strand Separation

Figure A.12: (a) The binding of the steroid molecule onto the formed DNA complex to induce strand separation. (b) The electrostatic potential of the binding site, which is achieved through hydrogen bonding at both ends, attaching to the phosphodiester coupling. (c) The formation of the transcription bubble on either side of the binding point up to two hundred nucleotides. The quality of the binding point will determine the width of the transcribing separation of strands.

sites, or longer chains of the weaker CG interactions.

It is noted that with ionic bonding from a positive two charge ion, there is a bijective code associated with the orientation of the steroid hormone type to the location of the bonding point, as indicated in Table A.1. This was originally described in my preprint [23].

Base Pair	**Steroid Orientation**	**Steroid Class**
A–T	$\mathcal{F}$	$\mathbb{H}$
T–A	$\mathcal{B}$	$\mathbb{H}$
C–G	$\mathcal{B}$	$\mathbb{S}$
G–C	$\mathcal{F}$	$\mathbb{S}$

Table A.1: A mapping of the location of the binding steroid to DNA nucleotide pair in conjunction with positive two ion.

A.3 Error correction

In developing error correction code and strategy, the solution is effective if the original code can be reconstructed, and then the status can be compared in corrected. For DNA encoding, it is possible to reconstruct the original encoded sequence of the orientation of the steroid molecules in the co-enzymatic complex, by knowledge of a fixed reference point and the nucleotide sequence. Having the steroid molecules available and serve as a matching template, error correction strategies are enabled. This is useful in either completely correcting for a damaged component, or at least in improving the probability that the corrected construct matches the original.

In addition, with the access to the DNA nucleotides provided around the region defined by the binding site of the steroid molecules, error correction can be performed on the single strand of it also provides a template to determine if the synthesis steps were in error, or if there is a misalignment when the double helix was formed, or during its transmission. It will favor a self-correction by the steroid molecule providing an external force to strengthen the neighboring groups to force a realignment. Other forms of self-correction are possible, including:

- The steroid molecule forms a redundancy check since both the DNA nucleotides and the steroid molecules were produced in a co-enzymatic way from the same molecular complex. As one influenced the other during their synthesis, the steroid molecule, in addition to other functions, also contributes in error correction since it can be used in recreating the original code.

- Error correction occurs through the self organization to achieve an interleaved group of structures, in which the binding of the mid-section hydroxyl group to the thymine molecule, or when coding through its association with the non-functional group of adenine, and thus not with the amine group of guanine. When the steroid is not favorably placed, for example guanine is where adenine should be, there is a natural affinity to displace the steroid and replace it with the one that can form the hydrogen bond.

- Error correction when the DNA is not separated, before the coding begins by matching the steroid connection to the pair. To determine which one is incorrect, since it is possible to recreate the original code and narrow the selection as to which one is incorrect, there is improved probability of picking and correcting the appropriate nucleotide. For example, if the pairs are mismatched A-G, and implementing the original code the possible choices in reading forward and backward that the incorrect matching should either be an A or a C, it will be known that G is in error, and A is the correct choice.

- Prior to translation when populating the strand, there is further error correction by the matching of the steroid structure to the complementary base pair. If a wall is not in place, the steroid molecule, which also forms the ceiling, will not be able to form an attachment point. Since the housing is not constructed, amino acid assembly will not proceed until the structure is completed.

The specific error correction procedures are indicated in Figure 8.4, which involve a replication of the missing components. The steroid molecules fill in the missing sections, matched to the available base pair. The phosphodiester backbone is then formed, and the steroid molecules are ejected as the previously damaged DNA section reverts to its double helix format. This procedure can also be performed with a single nucleotide missing. If the base pair is present but is damaged or mismatched, then it is conceivable of how an error correction would be performed through the insertion of a steroid molecule to stabilize the correct base pair, while a new nucleotide is formed.

A.4 Interaction Vessel

The interaction vessel example is indicated in Figure A.13, for a molecular model of the TGATGA sequence, which has the amino acids threonine. It shows that the amino acid binds onto both the steroid molecule at the ceiling, and the steroid molecule comprising the walls. Threonine is in position to directly interact with the second nucleotide comprising the triplet, which is guanine. The spacing of the interaction vessel is appropriate as the length of the steroid molecule is equal to twice the spacing from one wall to the next, and thus can cover two vessel openings and thereby retain only one opening for the insertion of the amino acid. The length of the amino acid from carboxyl group to amine group is approximately equal to the opening one vessel, and thus the combination with the steroid will establish the code of three nucleotide to one amino acid observed for the translation map of nucleotide sequence to amino acid.

The entire structure is maintained through hydrogen bonding. In Figure A.14 the van der Waals spheres are indicated to shown the integration of the overall structure. It is shown that the steroid molecule comprising the ceiling hydrogen bonds onto the third wall, that is the steroid molecule associated with the third nucleotide. The amino acid then connects the steroid molecule of the ceiling to the wall. It also bridges the gap and interacts with the base.

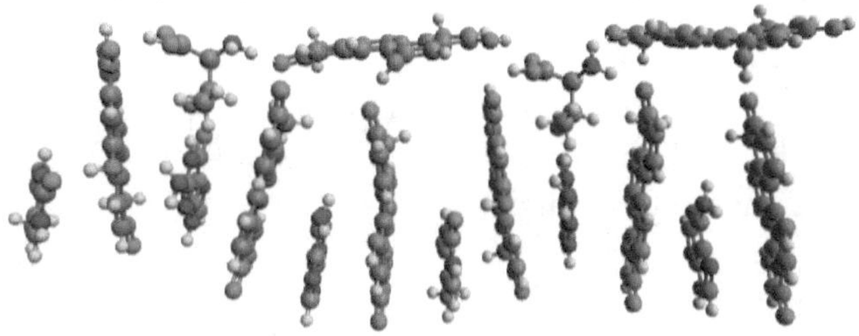

Figure A.13: For the assembly of the interaction vessel for the TGA sequence with the amino acid threonine, repeated twice, the molecular model indicates feasibility.

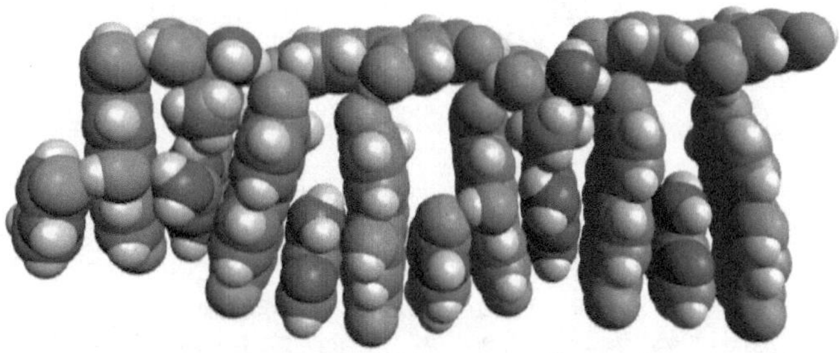

Figure A.14: The van der Waals representation of the TGATGA sequence, with the amino acid threonine is indicated.

A.5 Ionic binding

The role of ionic binding was examined in my initial work identifying a method of transcription based on direct binding of the steroid molecule to the DNA nucleotide pair [21], which interacts with an ionized phosphate group of the phosphodiester backbone. This bonding mechanism is useful because it enables a stronger connection and will result for a positive two charged ion in the involvement of two phosphate groups, thereby increasing its potential stabilization. To indicate its usage with Mg^{2+} ions are selected, of which other ions can be utilized of a positive two charge to link with adjacent phosphate groups, and include Ca^{2+}, Mn^{2+} and Zn^{2+}. In Figure A.15(b), the intermolecular hydrogen bonding between the steroid molecule and the thymine-adenine pairing is indicated through examination of the electrostatic potential. The intermolecular hydrogen bonding between the steroid molecule is indicated, which is 1.827 Å. The ionic binding is uniform, with the distance between the ion to the steroid group, for example the ketone group is 1.972 Å and the oxygen element of the phosphodiester backbone is 1.699 Å and 1.673 Å, as indicated in A.15(c) and (d). For the simulation results, the energy of the complex prior to binding was -1,161 kJ/mol, and after binding, it was -9,010 kJ/mol, thus providing for 7,849 kJ/mol of energy to dissipate through bond breaking.

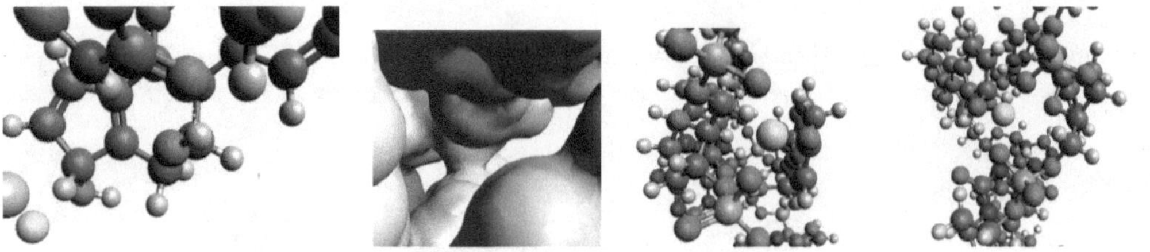

(a) Internal Hydrogen Bond (b) Electrostatic Potential (c) Ionic Binding (L) (d) Ionic Binding (R)

Figure A.15: (a) Intermolecular hydrogen bonding distances are achieved by the steroid to thymine coupling, as well as the internal thymine to adenine hydrogen bonding. (b) The electrostatic potential of the intermolecular hydrogen bond between the steroid triggering molecule and thymine is indicated, along with the internal two hydrogen bonds of adenine to thymine. (c) Ionic coupling of the end group of the steroid molecule, which will actually be a steroid hormone is achieved. (d) The two end groups of the steroid molecule are ionically coupled to the nucleotide pairs through adjacent phosphodiester groups.

In addition, it is noted that ionic bonding of the phosphate chain using an ion with a positive one charge, such as sodium, can be used to secure the steroid molecule to the DNA strands. The ketone end groups will attract towards the ions, along with the oxygen groups on the phosphate chain. If the phosphate chain has a hydroxyl group instead of its ionized form, then the hydroxyl group can induce hydrogen bonding with the ketone end group of the steroid molecule. In a similar fashion, when the steroid molecule has one hydroxyl end group on the five membered end group, which is the case for the start of the intermolecular coupling, and is the present, hydrogen bonding can be achieved with the ketone group associated with the phosphate. Ionic binding with a positive one charge enables charges at each phosphate coupling since the distance between each group is sufficient to prevent ionic repulsion. A positive two charge will allow a minimum of placements at every other phosphate group due to ionic repulsion.

A.6 Feedback control

In addition to providing the mechanisms for generating proteins from amino acids, the steroid molecule, as an intermediary, has to have an ability to participate in a feedback loop to promote those amino acid sequences which are more likely to be of use in the surrounding environment, and thus induce increases in the availability of the molecule to further participate in molecular interactions. In Figure A.16, the feedback loop is presented, in which the steroid molecule binding onto the strands induces strand separation, thereby accommodating transcription of amino acids into proteins. If the proteins are utilized, it will exit the strand, and thus the steroid structures are available for further binding. This feedback will continue onto the molecular structures inducing separation as well, as the forces from repeated implementation of the amino acid groupings on the strands may cause the degradation of the steroid strand separation element, and thereby terminate the strand separation and transcription. Therefore, proteins that feedback to maintain or inhibit strand separation will also have an impact on the transcription rates.

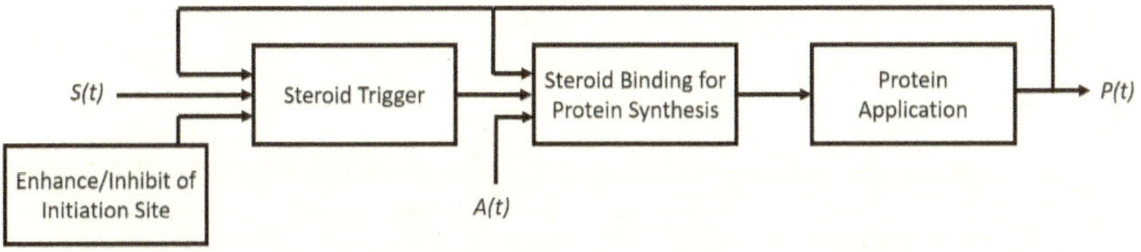

Figure A.16: The feedback loop to increase the rate of protein generation for those proteins that are more successful, that is find greater usage. In this block diagram, the steroid molecule concentration triggers the strand separation, which is also a function of the enhancer and inhibitor concentration which may influence the trigger zone. The protein molecule will also influence the steroid trigger, as a higher degree of use will place more force on the trigger point, and thus may disturb transcription. In the figure, $S(t)$ denotes steroid structures capable of inducing transcription, $A(t)$ denotes amino acid concentration over time, and $P(t)$ denotes proteins that are synthesized and diffuse away from the DNA molecule for usage in the surrounding environment.

As mediated through the steroid molecule, which functions to initiate transcription and as a workhorse for translation to enable protein generation, its regulation becomes a point at which cellular and later systems integration can take place. Control of the concentration and transportation of steroid molecules into the cytoplasm and nucleus thus are able to differentiate the cellular objective, and therefore is itself a building block for further development ultimately resultant in systems performance.

For example, the importance of steroid hormones is well known to physiology, and thus this map can be implemented at the nuclear level with direct binding to DNA. The intermediary of the glucocorticoid receptor acts as a transport mechanism to assist with the binding, but does not actual bind to the DNA chain like the steroid hormone. Control of the steroid hormone concentration is thus crucial to proper autonomic function, and thus external corticosteroids can provide benefit. Consequently, a unified description of systems performance is achievable through the steroid molecule, and the processes that impact it.

Appendix B

Methods

- Molecular modeling software: The software program Avogadro was utilized to build the three dimensional molecular models. The Avogadro software program was used to position the elements and calculate the bond distances. An optimization routine based upon steepest descent was implemented to position the molecules and the elements automatically. Matlab was used to generate the energy dissipation requirements when the DNA nucleotides were connected by the steroid molecule. Matlab was also used to produce the results used in the example of twenty five nucleotides, and twenty five steroid molecules in an enzymatic complex.

- Development of molecular models: The DNA nucleotides and steroid molecules were developed and synthesized cooperatively in an intermolecular manner by stacking pairs. The intermolecular hydrogen bonds were configured, and then the optimization software was used to orient the molecules, positioning to minimize energy. After establishing the stack, the phosphodiester backbone was assembled starting from the bottom at molecule 1, and then linking 3, 5, 7 and 9 by forming the N-C-O coupling, which will ultimately correspond to the nucleotide to sugar link. These results were performed in conjunction with the synthesis scheme described in the Appendix. After coupling, the sugar molecule was formed. Both sides were completed, with hydrogen bonding stabilizing the steroid triggering molecule to the phosphodiester structure. The internal nitrogen elements of the purine and pyrimidine components of the nucleotide pairs were then positioned accordingly, and the bonds were broken with the addition of oxygen groups selected at the appropriate position on thymine, cytosine, and guanine. The rotation was performed on the nucleotide pairs according to the codes, and ejected the pyrimidine molecules for this example, so as to eject the co-synthesized steroid molecules. Transcriptional positioning was performed by inserting the resultant steroid into the appropriate slot and calculating the energy reduction from the stabilization. For the ionic study, two Mg^{2+} ions were also included.

- Experimental: To obtain the relative activity of the synthetic steroids and steroid hormones, the literature was surveyed and representative values were selected for the activity of prednisone, prednisolone, cortisol, and dexamethasone [34]. The experimental structures of these steroid hormones were then overlaid to the experimental structures of the A-T nucleotide pair to achieve a third hydrogen bond by intermolecular coupling.

- Overall Procedure (journey): The development of this approach took the path of a top-down investigation. The initiating point was a mathematical analysis of a coordination of the nervous and cardiovascular systems to characterize a fever induced by pneumococcal vaccination in conjunction with exercise [35]. An investigation into the chemistry of the interaction resultant in the fever led to the concept of inhibition of cortisol by prostaglandins, which also led to the concept of coupling two Ca^{2+} ions at the glucocorticoid receptor to explain the differences in activity between the two molecules [36]. In further pursuit of a fundamental understanding of a fever resulted in transporting the glucocorticoid receptor to the nucleus to evaluate possible transcription of cortisol and prostaglandins through ionic binding directly on the DNA nucleotides. An evaluation of the structural symmetry of cortisol and the adenine-thymine base pair was determined, noting the intermolecular binding. Thus the symmetry of steroid molecules and DNA nucleotides was discovered [21, 22, 23]. This structural symmetry of steroid molecules and DNA nucleotides then led to the current results to define a basis for the transcription results perceived of cortisol on adenine-thymine, and other potential interactions between RNA nucleotide and other functional cyclic compounds. Hence, the original mathematical basis describing the systems and cellular responses are consistent with the nuclear developments, and thus enable an integrated solution.

Appendix C

Encoding sequence for DNA - Steroid complex in example

In Figure C.1, the encoded sequence that was used in the example is presented. It is comprised of a set of 25 pairs, hence 50 total, of steroid molecules. This encoded set was specified by a third party, which was a random number with uniform distribution. It results in the nucleotide sequence GAAATAGCACTGTCTCGTCGGTTGG with the first molecule of the sequence considered as the first nucleotide. For the first molecule since the preceding steroid structure was not specified it was consider that the last steroid molecule would determine if hydrogen bonding would be present.

It is noted that this sequence can be produced easily if the nucleotide sequence had been given instead. It would start from a G arrangement, which would be either PFBK or PBFK depending on the sequence of a preceding steroid molecule, which for this problem was considered as the fifiethmolecule, KFBP; thus the first molecule would be PFBK, since it would not have hydrogen bonding, and thus there is only one choice from Table 3.2. This corresponds to the steroid orientation KFBP. Thus, to achieve the next nucleotide, A, would involve selecting the appropriate sequence using Table 3.2. It requires hydrogen bonding since A-T only has two internal hydrogen bonds and three are required, and A is a pyrimidine, thus it must start with P. Thus, the choice is PBFK. This process is repeated until the sequence is finished.

P	F	B	K
K	F	B	P
P	B	F	K
K	B	F	P
P	F	B	K
K	F	B	P
P	B	F	K
K	B	F	P
K	B	F	P
P	B	F	K
P	B	F	K
K	B	F	P
P	B	F	K
K	B	F	P
K	F	B	P
P	F	B	K
P	F	B	K
K	F	B	P
K	B	F	P
P	B	F	K
K	F	B	P
P	F	B	K
P	B	F	K
K	B	F	P
K	B	F	P
P	B	F	K
K	B	F	P
P	B	F	K
K	F	B	P
P	F	B	K
K	F	B	P
P	F	B	K
P	B	F	K
K	B	F	P
K	B	F	P
P	B	F	K
K	B	F	P
P	B	F	K
P	F	B	K
K	F	B	P
P	F	B	K
K	F	B	P
K	F	B	P
P	F	B	K
K	B	F	P
P	B	F	K
P	F	B	K
K	F	B	P
P	F	B	K
K	F	B	P

Figure C.1: Encoding sequence for the example in encoding, transmitting, decoding and translating. To take a neutral example, The sequence of pairs was generated from a third party, that is uniform distributed random number generator repeated 25 times to produce an encoded sequence of a complex of 50.

Appendix D

Essays

In developing a consistent analytic approach to explain the origin of life, one of the benefits is the first shot at interpreting the results from a larger philosophical context. In this appendix, I present two essays connected to the developments of this book. The first essay relates to the aspects of the spirit, and the second essay relates to the aspects of the living, the here and now. Both essays are based on the fundamental molecular constructs comprising life function, namely the unified complex comprised ultimately of DNA nucleotides and steroid molecules. Both essays are written in a personal context based upon my life experiences. Growing up as a Catholic, the first essay is addressed towards the writings in the Bible, and drawing a correlation to the molecular results characterizing the origin of life. The second essay describes a life balance, in which the mental component must be aligned with the physical, which is meaningful based on my experiences in physical demanding competitions.

D.1 Affirming Biblical Writings with the Origin of the Genetic Code

Starting from functional constructs are generated and transmitted through a master pattern of single-use, there are numerous striking correlations of the new origin of life theory with the Biblical teachings. Such research into the origins of life inevitably, whether as the motivation or not, become entangled with philosophical analysis of religious beliefs, and thus, in developing a new framework for study of the first life function, it is appropriate to assess its characteristics with those from spiritual belief. It may impact the thought processes of the general population, especially in the present, and thus must consider the consistency with the teachings pursuant to tenets of religious writings. To initiate this exploratory study, here I assess the new results with regard to those of the Bible, particularly the Book of Genesis, to show significant correlation in support of the affirmation of the Biblical Code and scientific analysis.

In relationship to doctrines, Table D.1 compares beliefs according to a biblical interpretation and the scientific analysis presented in this book. It demonstrates agreement across several levels: the image of God and a master image for life function, for example, as a first pass comparison. For the scientific analysis, all life forms were derived from a master image, and whether the master image corresponding to that of humans, I continue my analysis to determine the differentiating factors between the forms of life, whether plant, birds, fish, mammals. And this would include just how the original master image was defined, the image of God, which from a scientific perspective would allow for the unified complex that would give rise directly to human. These are important concerns, and beyond that, there is strong correlation.

Based upon a scientific analysis, it would seem to be more consistent that the variety of life forms, since it came from an instantaneous creation of life function, were derived from specific events, rather than from gradual evolutionary changes. This would be consistent with the teachings in which life function was a separate and distinct moment, which did not necessarily require a long development period. Instantaneous life function is implied within the biblical teachings through a creation event, consistent with the notion that there was a moment, similar to a big bang start. From a scientific development, upon separation of the unified complex, it releases structure, through the double helix containing information content, and function, through the steroid molecule, such that immediate actions can start taking place. Amino acids are assembled and are brought into contact to produce proteins. Moreover, genetic rearrangements take place which creates the variety needed to produce all that is required for a variety of life.

Further, pertaining to a master image that is superior to that of the replicant, information content is much greater for the unified complex, than it is for the double helix. There is a significant reduction in releasing structure and function. And it defines a starting for which there really is no return, a one time event. The double helix then takes over and continues with replication, incorporated within a confined environment to utilize the life function elements of steroid molecule. There is a loss in generating structure from the double helix, which is lost when released from the double helix.

There are other areas of comparison, which would include the duality of good and evil. As indicated in the scientific analysis, it is possible to generate the same set of DNA nucleotides from the original master complex, however, that master complex would be completely out of phase. Thus the equivalent result is possible from a linear set of nucleotides, but the original complex contains an orientation that is opposite, for example PFBK versus PBFK, which would then need to be transmitted throughout the structure.

One item in need of clarification is the original structure of the unified complex. It has been shown that life is derived in the image of a master pattern. What then was the arrangement of the steroid molecules that would have proceeded the simultaneous release of structure and function to enable life function of the human form. The creation of the master image is still under evaluation. What is the master image minimal set to produce a unified function that can produce subsequent life function. Everything is known afterwards, and determining the initial set can be approached via an inverse mathematical problem.

The most basic result is that there is structure and there is function to life function, as created through the unified complex, the original design mechanism. It is taught then that life is more than just mental thought, that there is action and function. As the consistent in the ability to do both, the correlation of scientific analysis and biblical teachings should enable a consistent framework of spiritual development, based in scientific meaning.

Tenet	**Biblical**	**Scientific**
Creation	Image of God	Master Image
Replication	One time	One time
Pairing	Pairs	Pairs
Life function	Instantaneous	Instantaneous
Information Content	Structure and Function	Structure and Function
Human Evolution	Separate Path	TBD/separate paths
Duality	Good and Evil	Two opposite orientations
Purpose	Thoughts and action	Information and Access
Materials	Best materials	Nucleic Acids and Steroids

Table D.1: A table comparing the biblical teachings with those of the scientific developments presented in this book.

D.2 Balancing Mental & Physical Life Objectives

The origins of life function indicate that the true balance of life involves mental and physical acuity. From the unified complex, the DNA double helix and steroid molecules are derived so as to impart a life function at the onset of its release. Thus, there requires an informational content of DNA, which is akin to a mental pursuit, in combination with a physical acts of the steroid molecule to access and make use of the mental component. As life was derived from the combination of mental and physical affiliated processes, the true central purpose of life is one of balance, of which the demonstration of mental and physical accomplishments should result in a harmonious relationship with the original design, and lead to fulfillment.

While the DNA double helix has a prominent position in the study of molecular biology, it is but half of the original molecular complex from which life function sprang. It was co-synthesized in a format that was of a linear arrangement that itself was used as a template to shape the molecules that would ultimately allow access to the information. In rotating to the position of a double helix, enabled by a drive to a lower energy state, the DNA double helix allowed the transmission of information, stabilizing it for archival purposes. Effectively stashing an encoded message for archival purposes, as it contained the original code intended by the original design of the unified complex. But alone, the DNA molecule, and the mental processes of which it is considered, is not enough, as the information is available, but without access, it can not be put to use.

Thus there was the steroid molecule available, co-synthesized within the unified complex along with DNA molecules, such that when the complex was separated, immediate life function was induced, as the steroid molecule began immediately interacting with the DNA structure. The result was the synthesis of protein structures by designed and repeatable assembly of amino acids, enabled by the steroid construction of interaction vessels, comprising the wall and the ceiling, and in the delivery mechanism. The steroid molecule is the workhorse of inducing life function. It is required otherwise DNA is a useless molecule, which contains information, but without the mechanics to read it, serves little purpose. Moreover, the manner in which it is accessed, the areas and the length of time from which it is read, determines the result, and the ultimate function of the system.

Therefore, the true balance of life is one of intelligence and of physical capability, as both were defined in the original complex. The success of the individual should therefore be consistent of what it took to become a life function, both intelligence and physical capability aligned with the intelligence. Having one without the other is not in alignment with the original purpose of the unified complex. It is necessary to balance the mental and physical to promote an innate sense of fulfillment. While it seems interesting to promote one aspect over another, it is not in alignment with the true definition of the unified complex, the design of DNA and its purpose. Thus, it offers a multiplicative function, rather than an additive function, with the metric as $Y = MP$, where Y is the fulfillment factor, and M is the mental development and P is the physical capability, of which $M + P = 1$. Thus, the optimal point for maximizing the function is found by taking the partial derivative of Y with respect to M and P. It is interesting that this results in $M = 1/2$ and $P = 1/2$. Both are needed, and should be emphasized equally to achieve the best fulfillment in life. This is also consistent with the unified complex construct, which was comprised of DNA molecules indicating half M and steroid molecules indicating half P.

The ability to survive needs to take precedence to achieving a life balance, and this might require a shift in the proportion of M or P. If it is a financial struggle or the desire to prosper in a chosen profession or educational endeavor, there may be the need to shift towards an M of one, with little regard for P. Likewise, if there is a physical pursuit, say a competition of athletic endeavor, a shift of P to one may be required. But, this should only be done for a period of time, and a shift towards a central balance should be the driving factor. Its purpose in shifting away from center, should be to stabilize the ability to get back to center in a better position than previously. More stable, and at a higher level of growth, so as to avoid decay.

Derived from a unified complex of DNA nucleotides and steroid molecules, there is an inherent capacity for mental and physical conquests. To align ones life towards a balance of each, will enable a consistency that is implicit to the creation of life function. Therefore, one should strive for a balanced pursuit of intelligence and physical strength, without neglecting one for the other, for central agreement with the fundamental design.

Appendix E

Statement of No Conflicts of Interest

The author asserts no conflicts of interest. No external funding was used for this study.

Bibliography

[1] F. Dyson, *Origins of life.* Cambridge University Press, 1999.

[2] A. Lazcano and S. L. Miller, "The origin and early evolution of life: prebiotic chemistry, the pre-rna world, and time," *Cell*, vol. 85, no. 6, pp. 793–798, 1996.

[3] J. D. Sutherland, "Opinion: Studies on the origin of life—the end of the beginning," *Nature Reviews Chemistry*, vol. 1, no. 2, pp. 1–7, 2017.

[4] R. Cojocaru and P. J. Unrau, "Origin of life: Transitioning to dna genomes in an rna world," *eLife*, vol. 6, p. e32330, 2017.

[5] P. Forterre, J. Filee, and H. Myllykallio, "Origin and evolution of dna and dna replication machineries," *Madame Curie Bioscience Database [Internet]. Landes Bioscience; 2000-2013.*, 2013, https://www.ncbi.nlm.nih.gov/books/NBK6360/.

[6] W. Gilbert, "Origin of life: the rna world," *Nature*, vol. 319, no. 20, p. 618, 1986.

[7] H. M. Goodman and A. Rich, "Formation of a dna-soluble rna hybrid and its relation to the origin, evolution, and degeneracy of soluble rna," *Proceedings of the National Academy of Sciences of the United States of America*, vol. 48, no. 12, p. 2101, 1962.

[8] S. L. Miller and J. L. Bada, "Submarine hot springs and the origin of life," *Nature*, vol. 334, no. 6183, pp. 609–611, 1988.

[9] S. Nakashima, Y. Kebukawa, N. Kitadai, M. Igisu, and N. Matsuoka, "Geochemistry and the origin of life: From extraterrestrial processes, chemical evolution on earth, fossilized life's records, to natures of the extant life," *Life*, vol. 8, no. 4, p. 39, 2018.

[10] K. A. Lanier and L. D. Williams, "The origin of life: models and data," *Journal of molecular evolution*, vol. 84, no. 2-3, pp. 85–92, 2017.

[11] S. Hoshika, N. A. Leal, M.-J. Kim, M.-S. Kim, N. B. Karalkar, H.-J. Kim, A. M. Bates, N. E. Watkins, H. A. SantaLucia, A. J. Meyer *et al.*, "Hachimoji dna and rna: A genetic system with eight building blocks," *Science*, vol. 363, no. 6429, pp. 884–887, 2019.

[12] R. Service, "Researchers may have solved origin-of-life conundrum," *Science*, 2015, doi:10.1126/science.aab0325.

[13] B. H. Patel, C. Percivalle, D. J. Ritson, C. D. Duffy, and J. D. Sutherland, "Common origins of rna, protein and lipid precursors in a cyanosulfidic protometabolism," *Nature chemistry*, vol. 7, no. 4, p. 301, 2015.

[14] M. Preiner, S. Asche, S. Becker, H. C. Betts, A. Boniface, E. Camprubi, K. Chandru, V. Erastova, S. G. Garg, N. Khawaja *et al.*, "The future of origin of life research: bridging decades-old divisions," *Life*, vol. 10, no. 3, p. 20, 2020.

[15] P. Marshall, *Evolution 2.0: Breaking the Deadlock Between Darwin and Design.* BenBella Books, Inc., 2015.

[16] Á. Kun and Á. Radványi, "The evolution of the genetic code: Impasses and challenges," *Biosystems*, vol. 164, pp. 217–225, 2018.

[17] E. V. Koonin and A. S. Novozhilov, "Origin and evolution of the universal genetic code," *Annual review of genetics*, vol. 51, pp. 45–62, 2017.

[18] R. Schrodinger and E. Schrödinger, *What is life?: With mind and matter and autobiographical sketches.* Cambridge University Press, 1992.

[19] J. D. Watson and F. H. Crick, "Molecular Structure of Nucleic Acids," *Nature*, vol. 171, no. 4356, pp. 737–738, 1953.

[20] J. Monod, "On chance and necessity," in *Studies in the Philosophy of Biology.* Springer, 1974, pp. 357–375.

[21] C. Schaper, "Endogenous Binding of Steroid Molecules to DNA Nucleotides by a Ca2+/PO4- Process to Enable Gene Transcription," *ChemRxiv*, 2020, https://doi.org/10.26434/chemrxiv.11868261.v2.

[22] ——, "Structural Symmetry of DNA Nucleotides and Steroid Hormones," *ChemRxiv*, 2020, https://doi.org/10.26434/chemrxiv.11991567.v1.

[23] ——, "New Code for DNA Nucleotide Sequences," *ChemRxiv*, 2020, https://doi.org/10.26434/chemrxiv.12003786.v1.

[24] D. Becker, "Basic and clinical pharmacology of glucocorticosteroids," *Anesthesia Progress*, vol. 60, no. 1, pp. 25–32, 2013.

[25] C. Schaper, "Intermolecular Binding of Functional Cyclic Compounds to RNA Nucleotides by an Ionic Process," *Preprints*, 2020, https://doi.org/10.20944/preprints202004.0522.v1.

[26] N. Hoffmann, "Photochemical reactions as key steps in organic synthesis," *Chemical Reviews*, vol. 108, no. 3, pp. 1052–1103, 2008.

[27] U. Reuther and A. Hirsch, "Synthesis, properties and chemistry of aza [60] fullerene," *Carbon*, vol. 38, no. 11-12, pp. 1539–1549, 2000.

[28] O. Vostrowsky and A. Hirsch, "Heterofullerenes," *Chemical reviews*, vol. 106, no. 12, pp. 5191–5207, 2006.

[29] Y. Li, G. Zhang, D. Wang, B. Xu, D. Xu, N. Lou, and L. Gan, "Fullerene-based macro-heterocycle prepared through selective incorporation of three n and two o atoms into c60," *Angewandte Chemie International Edition*, vol. 55, no. 47, pp. 14 590–14 594, 2016.

[30] J. B. Siegel, A. Zanghellini, H. M. Lovick, G. Kiss, A. R. Lambert, J. L. S. Clair, J. L. Gallaher, D. Hilvert, M. H. Gelb, B. L. Stoddard *et al.*, "Computational design of an enzyme catalyst for a stereoselective bimolecular diels-alder reaction," *Science*, vol. 329, no. 5989, pp. 309–313, 2010.

[31] A. Sanyal and J. K. Snyder, "Stereoselective diels- alder reactions of chiral anthracenes," *Organic letters*, vol. 2, no. 16, pp. 2527–2530, 2000.

[32] S. E. Wheeler, A. J. McNeil, P. Muller, T. M. Swager, and K. Houk, "Probing substituent effects in aryl- aryl interactions using stereoselective diels- alder cycloadditions," *Journal of the American Chemical Society*, vol. 132, no. 10, pp. 3304–3311, 2010.

[33] K.-C. Chou, "Low-frequency vibrations of dna molecules," *Biochemical Journal*, vol. 221, no. 1, pp. 27–31, 1984.

[34] J. Labrin and K. Cunningham, "What corticosteroid is most appropriate for treating acute exacerbations of copd," *The Hospitalist*, June 2011.

[35] C. Schaper, "Dynamics and Internal Control of Body Temperature in Response to Infectious Agents and Other Causal Factors," *bioRxiv*, 2019, https://doi.org/10.1101/566679v1.

[36] ——, "Competitive Inhibition of Cortisol by Prostaglandins at the Ligand Binding Domain of Glucocorticoid Receptors," *bioRxiv*, 2019, https://doi.org/10.1101/851501.

Design of DNA, Genetic Codes, and Life Function

Charles D. Schaper, Ph.D.

The origin of life is one of the greatest mysteries. The mechanism for the synthesis of DNA is synonymous with the chemical origin of life, and theories have been developed along many lines of reasoning, but resolving all requirements remains a challenge, such as defining an objective path to produce sequences of encoded nucleotides paired as adenine to thymine and guanine to cytosine. Here, original research is presented for the design of DNA, genetic codes, and life function. The basis for this work is the discovery of a molecular primer evident in the structural symmetry between DNA nucleotides, steroid molecules, and steroid hormones. Following this discovery, the research introduces a new concept of synthesizing both structural and functional characteristics of DNA at the same time within a single unified complex of interleaved steroid molecules which form reaction vessels that serve as co-enzymatic building blocks. The establishment of the DNA nucleotide code is among the very first synthesis steps, thereby inducing a digital communications channel of encoding, transmission, decoding, translation, including error correction and replication. Moreover, as a consequence of the intermolecular synthesis of both structure and function within a unified complex, there is a culminating process step that sets forth in motion both DNA and the steroid structures that subsequently control translation and replication, thereby resulting in the instantaneous release of life function.

Providing a solution to the question of the origin of DNA and life function, the book describes evidence of structural symmetry in DNA nucleotides of steroid molecules as prove that DNA is derived from a complex of higher information content, created in the image of a master pattern, that imparts not only structure, in the form of double helix DNA, but also function, in the form of steroid molecules. In developing this strategy, Dr. Schaper shows: The development of molecular information systems; The translation code of life mapping nucleotide triplets to amino acids is arranged; The requirements to impart structure and function within a unified complex; The mechanisms for the origin of nucleic acid replication and error correction. Those who want a scientific basis for the origin of life function will need to read this book!

Molecular Sciences Publishing House

www.ingramcontent.com/pod-product-compliance
Lightning Source LLC
LaVergne TN
LVHW102231300926
849160LV00006B/94
* 9 7 8 1 7 3 5 3 7 2 1 3 6 *